The ICE Design and Construct Contract

A Commentary

The ICE Design and Construct Contract

A Commentary

Brian Eggleston

CEng, FICE, FIStructE, FCIArb

OXFORD

BLACKWELL SCIENTIFIC PUBLICATIONS

LONDON EDINBURGH BOSTON

MELBOURNE PARIS BERLIN VIENNA

Editorial Offices:
Osney Mead, Oxford OX2 0EL
25 John Street, London WC1N 2BL
23 Ainslie Place, Edinburgh EH3 6AJ
238 Main Street, Cambridge,
Massachusetts 02142, USA
54 University Street, Carlton,
Victoria 3053, Australia

Other Editorial Offices:
Librairie Arnette SA
1, rue de Lille
75007 Paris
France

Blackwell Wissenschafts-Verlag GmbH
Düsseldorfer Str. 38
D-10707 Berlin
Germany

Blackwell MZV
Feldgasse 13
A-1238 Wien
Austria

First published 1994

Set by DP Photosetting, Aylesbury, Bucks
Printed and bound in Great Britain by
Hartnolls Ltd, Bodmin, Cornwall

DISTRIBUTORS

Marston Book Services Ltd
PO Box 87
Oxford OX2 0DT
(*Orders:* Tel: 0865 791155
Fax: 0865 791927
Telex: 837515)

USA
Blackwell Scientific Publications, Inc.
238 Main Street
Cambridge, MA 02142
(*Orders:* Tel: 800 759-6102
617 876-7000)

Canada
Oxford University Press
70 Wynford Drive
Don Mills
Ontario M3C 1J9
(*Orders:* Tel: (416) 441-2941)

Australia
Blackwell Scientific Publications Pty Ltd
54 University Street
Carlton, Victoria 3053
(*Orders:* Tel: 03 347-5552)

British Library
Cataloguing in Publication Data

A catalogue record for this book is available from the British Library

ISBN 0–632–03697–4

Library of Congress
Cataloging in Publication Data

Eggleston, Brian, CEng.
The ICE design and construct contract: a commentary / Brian Eggleston.
p. cm.
Includes bibliographical references and index.
ISBN 0–632–03697–4
1. Civil engineering contracts–Great Britain. 2. Institute of Civil Engineers (Great Britain). Conditions of contract and forms of tender, agreement, and bond for use in connection with works of civil engineering construction. I. Title.
KD1641.E336 1994
343.41′078624–dc20
[344.10378624] 93-41495
CIP

Contents

Preface

No one who spent the 1980s as I did, with one foot in the building industry and the other in civil engineering, could have doubted that the remarkable growth of design and build contracting in the building industry in that decade would be followed by a similar trend in civil engineering in the 1990s.

The question was, would a standard form of contract emerge to provide the impetus in civil engineering which JCT Design and Build Form (1981) provided in the building industry?

In the event, not one, but a variety of forms has emerged. However it remains to be seen which of these forms will provide the major stimulus to the growth of design and construct contracting in civil engineering.

As a lifelong user of the ICE Conditions of Contract it is my hope that the ICE Design and Construct Conditions of Contract will lead the way. And my purpose in writing this book so soon after the introduction of the Design and Construct Conditions is essentially to promote their use. The fact that I query the clarity of some of the provisions and recommend amendments to some of the clauses is no more than detail. My overall view is that the Design and Construct Conditions are generally well drafted and fair to both parties. They deserve to succeed and I think they will, not least because having their origins in traditional ICE Conditions of Contract they start with a unique advantage over other competing forms. The majority of users will feel immediately comfortable with the familiar style of wording and the retention of a well established clause numbering system. And a venture into design and construct contracting with these Conditions will not be wholly a journey into the unknown.

In connection with the production and promotion of the Design and Construct Conditions of Contract two names deserve a special mention, Mr Stuart Mustow and Mr Geoffrey Hawker. To these two gentlemen who have done so much to ensure that the ICE Conditions of Contract in various forms remain the basis of civil engineering contracts and who have been so helpful to me over the years I dedicate this book.

Brian Eggleston
5 Park View
Arrow
Alcester
Warwickshire

September 1993

Author's note

Phraseology

Rather than repeat throughout the book the full title of the Conditions – the ICE Design and Construct Conditions of Contract – I have resorted to abbreviations where appropriate.

Generally I refer either to the D and C Conditions or to the Conditions.

Capitals

As a matter of style capital letters have been used sparingly. The result, I hope, is easier reading. However, it does mean that defined contractual terms such as the employer and the contractor appear completely in lower case.

Text of the Design and Construct Conditions

For commercial reasons it has not been possible to include the text of the Design and Construct Conditions in this book.

I have assumed, therefore, that readers will have to hand a copy of the Conditions.

Chapter 1

Introduction

1.1 Publication

The first edition of the *ICE Design and Construct Conditions of Contract* was published in October 1992 by Thomas Telford Services Ltd for the joint sponsoring authorities:

- The Institution of Civil Engineers
- The Association of Consulting Engineers.
- The Federation of Civil Engineering Contractors

Drafting committee

The sponsoring authorities have for many years maintained a permanent joint committee – the Conditions of Contract Standing Joint Committee (CCSJC) – to draft new conditions of contract and to keep existing documents under review.

The Standing Joint Committee is responsible for traditional ICE conditions of contract and various specialist forms and prior to the Design and Construct Conditions of Contract its most recent works were the ICE Minor Works Conditions of Contract (1988) and the ICE Conditions of Contract Sixth edition (1991). The committee was able to draw on its experience in drafting traditional forms of contract in the preparation of the Design and Construct Conditions of Contract and the result is a form of contract which will feel immediately familiar to users of other ICE contracts.

Guidance notes

Until recently it was not the policy of CCSJC to issue guidance notes on its contracts. That policy has now changed, perhaps because most other standard forms of contract are accompanied by the draftsmen's notes for guidance.

Guidance notes on the *ICE Design and Construct Conditions of Contract* are now available from Thomas Telford from late 1993.

1.2 The growth of design and construct contracting

For civil engineers, design and construct contracting is not a new concept. The pioneers of the nineteenth century were frequently both the designers and the risk takers.

Package deals

But design and construct, and design and build, as known today did not really begin to develop in the UK until the late 1950s when some contractors began to offer package deals. Such deals gave the employer the benefit of single point responsibility avoiding the need for the employer to enter into separate agreements with designers and contractors.

Although package deals appealed to certain industrialists, employers generally in the construction industry were slow to move away from the traditional arrangement of employer's design. In part this may have been because codes of professional conduct prevented some designers from being company directors but it probably had more to do with the fact that many employers, especially in the public sector, had their own in-house design teams. So even as late as 1975, Lord Denning MR in the *Greaves* case, discussed in Chapter 4, was able to describe a package deal as a new kind of building contract.

Design and build

By the early 1980s, however, a new mood of commercial awareness was awake in the UK and traditional arrangements of all kinds were being examined for efficiency and value for money. In the building industry, industrialists and developers were the first to recognise the potential advantages of rapid procurement times, fixed prices and single point responsibility of design and build. And within a decade design and build had spread from being a novelty to the normal; so that by the end of the 1980s as much as 40% of all new building work was being procured by the design and build system.

JCT 81

Much of the credit for the growth of design and build contracting in the 1980s can be attributed to the foresight of the building industry's Joint Contracts Tribunal which in 1981 published a version of its new standard form of traditional building contract (JCT 80) called the Standard Form of Building Contract with Contractor's Design (JCT 81).

JCT 81, because it was produced in consultation with all sides of the building industry, had the immediate effect of conferring respectability on a method of procurement which until then had been regarded by some as being not quite fully professional.

The new contract contemplated not only that professional design practices would undertake work for contractors but also that in some circumstances, where the employer had engaged firms to commence the design, those firms could be novated to the contractor to complete the design.

By the end of the 1980s some employers, including by then some from housing associations and public sector bodies, were so impressed by the trouble free nature of design and build contracting that contracts which had been wholly prepared under the traditional system were being let under JCT 81. Not all contractors regarded this distortion of design and build principles with enthusiasm but the practice illustrated the determination of employers to use forms of contract which they perceived to deal best with their needs.

Civil engineering – design and construct

For civil engineering, the movement towards design and construct is only now, in the early 1990s, gathering pace.

The absence of a dedicated design and construct form until the publication of the ICE Design and Construct Conditions of Contract has been one restraining factor. Another has been the belief that civil engineering unlike building provides little scope for the contractor to exercise originality in design – a belief of some truth in relation to straightforward roadworks and sewerage works but of little application elsewhere.

A third factor is that in many civil engineering projects the amount of site investigation to be undertaken and the amount of design work necessary to prepare the employer's requirements makes design and construct appear to be an impracticable procurement option.

But notwithstanding these difficulties employers are turning towards design and construct in civil engineering with the lead being taken by big spending employers in the water industry and road building. They, perhaps earlier than others, have recognised the financial benefits that design and construct can bring with its certainty of price and freedom from claims.

1.3 *Characteristics of design and construct contracting*

The essence of design and construct contracting is that the contractor carries responsibility for the design and for the performance of the designers as well as carrying his normal responsibilities for construction.

From this, three points follow of great importance to the employer. Firstly, that at common law the works as designed should be fit for their purpose. Secondly, that the contractor's price will normally be on a lump sum basis. Thirdly, that the contractor's opportunities for claims will be greatly reduced.

Fitness for purpose

The point on fitness for purpose is that the law likens design and construct contracting to the supply of goods. And at common law and under statute goods should be reasonably fit for the purpose for which they are intended.

However, if the employer engages his own designers all he receives from them is a service and their responsibility for design does not extend to fitness for purpose but is limited to using reasonable skill and care.

Lump sum pricing

Many employers would put certainty of price at the top of their list of priorities for contractual requirements. That certainty can rarely be achieved without lump sum pricing.

Design and construct contracts are inherently lump sum because the contractor takes responsibility for his own quantities. That represents a major shift of risk from the employer to the contractor compared with remeasurement type contracts and it is a major attraction to the employer.

Reduced opportunities for claims

The phrase used repeatedly in connection with design and construct contracting is that it offers single point responsibility.

The practical application of this is that the employer does not have the trouble of dealing with numerous firms. The financial benefit to the employer is that he is not liable to the contractor for the defaults of those firms when they are engaged directly by the contractor.

This immediately eliminates a major source of contractor's claims and it reinforces the certainty of price achieved by lump sum pricing.

Other potential benefits

- Speed of procurement. In some circumstances design and construct contracting can reduce the preparation time of a contract and lead to an earlier start on site. One reason is that the contractor can finalise the details of his design as the works proceed and less detail needs to be available at the time of tender.
- Choice in design. Competition in design produces choice for the employer and it encourages innovation. But to get the best out of competition the employer's requirements must not be so tightly drawn as to be inflexible.
- Buildability. In traditional contracting, contractors often complain of unworkable details or identify how with some variation the works could be better built. Design and construct contracting offers the contractor the opportunity to use his skills and expertise to produce buildable designs and that should be reflected in the prices available to the employer.

1.4 *Comment on the Design and Construct Conditions*

Background and origin

Unlike the New Engineering Contract, the ICE Design and Construct Conditions of Contract are neither novel in their drafting nor a radical departure from earlier forms of contract. The Conditions are quite clearly an adaption of the traditional ICE Conditions of Contract Sixth edition to accommodate the contractor's responsibility for design.

With one or two exceptions, the clause numbering of the ICE Sixth edition has been retained and so has much of the drafting.

Comparison with the ICE Sixth edition

In the explanatory memorandum released with the Design and Construct Conditions the drafting committee expresses the hope that users of the Conditions will not waste time with sterile comparisons with the ICE Sixth edition but will instead approach the new Conditions as an organic and internally consistent whole.

This approach has much to recommend it in the sense that legal construction of the Design and Construct Conditions will be made from the contract itself and not by reference to some other form of contract. And it is also the right approach in the sense that it is no use trying to compare apples with oranges. The two sets of conditions set out to convey different obligations and responsibilities.

However, what cannot be ignored is the fact that users of the Design and Construct Conditions will frequently have had years of experience using the traditional ICE Fifth and Sixth editions and they will find identical, or nearly identical clauses to those they know and understand now included in the Design and Construct Conditions.

To a great extent this can be seen as an advantage to the Design and Construct Conditions. It will certainly be a major selling point in their use. But it is also potentially a source of great danger because clauses, although identically worded, can take on one meaning in the context of one contract and another meaning in the context of some other contract.

For this reason, in this book, I have thought it appropriate to show for every clause of significance in the Conditions the implications of the transposition of the text from the ICE Sixth edition.

But I do additionally believe that it will help users of the Design and Construct Conditions to gain confidence and understanding in their use if they can relate the provisions to a familiar benchmark.

Responsibility for design

The aim of the Design and Construct Conditions is to make the contractor wholly responsible for design including that carried out by, or on behalf of, the employer. A secondary aim seems to be to place the contractor's responsibility on a skill and care basis thereby excluding from the Conditions any implied warranty on fitness for purpose of the design.

The first of these aims is broadly achieved although not without potential difficulty in respect of the employer's design and notwithstanding the enigmatic phrase which appears from place to place within the Conditions 'design for which the Contractor is not responsible under the Contract'.

As to the second aim on skill and care, doubts remain as to exactly what has been achieved in the drafting. Fitness for purpose may not have been excluded. Debate on this will run and run; but in a vote I would join the fitness for purpose 'not excluded' side.

Formation of the contract

The award process for design and construct contracting rarely achieves the simplicity of traditional contracting where a letter of acceptance of the contractor's tender is all that is necessary. Usually in design and construct there will be negotiations between the submission of the tender and the award to clarify the details of the contractor's submission and to agree what changes need to be made to either or both of the employer's requirements and the contractor's submission.

The Design and Construct Conditions recognise this in their definition of contractual terms but care still needs to be taken with the model forms of tender and agreement to avoid confusion in the formation of the contract. The form of tender itself still indicates simple acceptance and the form of agreement is not wholly consistent with the form of tender.

Definition of the contract

A potential problem of some scale in the Design and Construct Conditions lies in the definition of the 'Contract'.

Allowance has only been made for changes in the employer's requirements and it is far from clear how any changes the contractor might wish to make, or might have to make, in his submission, can be brought within the contract.

Obligations of the contractor

The primary obligation of the contractor under the Design and

Construct Conditions is to design, construct and complete the works in accordance with the contract. Since the Conditions leave no doubt that the employer's requirements take precedence over the contractor's submission that means, in effect, to design and construct in accordance with the employer's requirements.

This may seem on the face of it an eminently sensible arrangement. But it has its difficulties. The contractor may not be under a contractual obligation to provide anything he has offered in his submission which is in excess of the employer's requirements; particularly if the contractor can point to a discrepancy between the employer's requirements and his submission.

Impossibility

Impossibility may be absolute – in which case nothing can be done to resolve it – or it may be resolvable by change. Impossibility may also be judged from a practical commercial point of view as it was in the *Turriff* case discussed in Chapter 8. In construction contracts impossibility is rarely absolute.

In traditional ICE contracts non-absolute impossibility is a problem for the engineer to resolve. He is required to vary the works to ensure their completion. Consequently very few contracts are abandoned for impossibility.

In the Design and Construct Conditions there is no express obligation on the employer's representative to deal with impossibility – the burden, if it exists, appears to fall on the contractor. And the extent of that burden depends on how the proviso to the contractor's obligation to design and construct and complete the works 'save in so far as it is legally or physically impossible' is to be interpreted. On one view the proviso indicates a very heavy burden – that the contractor's performance is excused only by absolute impossibility. But on another view it has the opposite effect and it relieves the contractor of his obligation to complete the work for any non-absolute impossibility.

Unforeseen conditions

The Design and Construct Conditions retain the traditional provision for the employer to take the risk on unforeseen physical conditions. It is certainly debatable whether this is appropriate for design and construct contracts and it will not be surprising if many employers follow the practice being introduced by the Department of Transport of obtaining tender prices with and without the unforeseen conditions clause of their design and construct form.

Variations

In design and construct contracts it is usual to talk of changes rather than variations. This emphasises that either the employer can change his requirements or the contractor can change his proposals.

The Design and Construct Conditions retain the traditional 'variation' phraseology but they do so only in respect of changes to the employer's requirements.

The absence of any scheme in the Conditions for dealing with changes advanced by the contractor looks like a serious omission. It has repercussions on obligations and it has repercussions on price.

Also of concern is the absence of any provision for valuing omission variations. Users of the Conditions will need to rectify this.

Measurement

To retain flexibility in the use of the Design and Construct Conditions the draftsmen have retained, with some modification, some of the measurement provisions of traditional ICE contracts. These need to be read with care.

In many cases they will be of no effect; but even when they take effect their application will be different than in traditional contracting.

Payments

It may not be apparent from first reading, but the Design and Construct Conditions unlike traditional ICE contracts do not give the contractor an automatic right to monthly payments. The Conditions rely on rules for payment expressed elsewhere in the contract.

This has the great merit of flexibility because there are now many different schemes for interim payments in operation and the employer can choose the scheme he thinks most appropriate for his particular contract.

The payment scheme will have to be incorporated into the contract either through the employer's requirements or as an appendix/schedule. Along with the payment scheme or elsewhere should be a requirement on lump sum contracts for a contract sum analysis – a matter not mentioned in the Conditions but essential for payment and valuation purposes.

1.5 *Views on drafting*

It is some time since I have seen in my professional work as an arbitrator or contract consultant a standard form of contract which has not been amended. Not infrequently I am involved because of the amendments.

I hesitate therefore to recommend amendments to any standard form and when I do it is usually with the proviso that clients take legal advice before making any changes.

Nothing would have pleased me more than to have written this book with only praise for the drafting of the Design and Construct Conditions but the reality is that I am not comfortable with some of the clauses and I believe that some users of the Conditions will share my views.

To assist those with the task of putting contracts together and thinking of using the Design and Construct Conditions I have scheduled at the end of this book a list of the clauses on which I have reservations. The schedule gives the section reference where each matter is considered in detail.

However, I would repeat my point above that amendments to the Conditions should only be made with legal advice unless they are straightforward.

Chapter 2

The employer's requirements

2.1 *Introduction*

The foundation for success in design and construct contracting rests with the quality and clarity of the employer's requirements.

If the employer knows what he wants and he can convey that message to the contractor, preferably without change, he is entitled to expect the performance the contract demands. If the employer either does not know what he wants or he fails to impart to the contractor precisely what his requirements are then the design and construct procurement route is likely to be, at best, an unhappy path and, at worst, the road to disaster.

Responsibility for preparation

Design and construct contracting is sometimes seen by employers as a soft option – leave the contractor to sort everything out and avoid the costs of employing a professional project team. That rarely works; even in turnkey projects, so called because the employer just turns the key and walks into a fully finished and fitted out installation ready for operation. Whether the installation is a road bridge, a water treatment works or a factory for canning beans, the employer has to specify standards of work and/or standards of performance. He also needs to have the services of some professional advisers to draft his requirements, monitor progress, authorise payments, certify completion and/or acceptance and generally manage the project.

For civil engineering projects, employers will continue to rely on qualified engineers, whether in-house staff or external consultants, to provide the expertise needed to prepare the employer's requirements. Although there will be competition from other professions, the majority of employers are likely to appoint qualified engineers as the employer's representative.

Detail of the requirements

There is no standard code of practice which applies across the whole range of design and construct contracts as a guide to preparation of the employer's requirements. For a start the requirements are too diverse but in any event contractual provisions vary so widely in the forms of contract that a standard code would be inapplicable.

Nevertheless some standard forms of contract are accompanied by either guidance notes on the preparation of the employer's requirements (as JCT 81) or they are structured to include a specified set of schedules to cover the employer's requirements (as the IChemE forms).

The D and C Conditions as they stand offer little to indicate what should be provided in the employer's requirements to get the best out of the Conditions. The definition in clause 1(1)(e) simply refers to requirements describing 'standards performance and/or objectives'. Such flexibility has its advantages but it does mean that every employer embarking on a contract with the D and C Conditions has a unique problem.

Official guidance notes will be welcome if only to provide an outline pattern of the shape of the employer's requirements and a checklist of points for inclusion.

2.2 *Employer's requirements in the Design and Construct Conditions*

Definition

Clause 1(1)(e) defines the employer's requirements as the requirements which may be identified as such:

- at the date of the award of the contract, and
- any subsequent variations thereto

and which may describe:

- standards
- performance
- objectives

to be achieved by the works or parts thereof.

This contractual definition, although apparently very wide in its scope, is in some respects deficient. It fails to emphasise that the primary purpose of the requirements is to describe the works and their location; and it omits any reference to requirements which are to be placed on the contractor by way of contractual obligations – for example – to what extent, if any, a quality assurance system is to apply.

Variations

By definition, the employer's requirements include subsequent variations. This is a point of great importance. It means that whenever the phrase 'employer's requirements' is used in the Conditions it includes not only the requirements at the date of award of the contract but any subsequent changes. As will be seen in Chapter 3, this does not apply to the contractor's submission; and the contractual effects of that are potentially far reaching.

Under clause 51(1) only the employer's representative has power to vary the employer's requirements – and then only after consultation with the contractor. Such variations may include additions or omissions.

There is no express obligation under clause 51 for the employer's representative to order variations which are necessary to ensure completion of the works, as is the position with the engineer under the ICE Sixth edition. This is presumably because the contractor has responsibility for design of the works. But if completion in accordance with the employer's requirements is impossible (a matter discussed in later chapters) there may be an implied obligation on the employer's representative to order variations simply to give the contract business efficacy.

Discrepancies

Clause 5(1)(c) deals with ambiguities or discrepancies in the employer's requirements. They are to be explained or adjusted by the employer's representative with appropriate instructions in writing to the contractor.

If the instructions cause the contractor delay or extra cost he is entitled to an extension of time for completion and payment of such cost as is reasonable. An allowance for profit is made in respect of additional work.

Status of the employer's requirements

In the D and C Conditions the employer's requirements, unlike the position in some other standard forms, take precedence over the contractor's submission.

This emerges most clearly from clause 5(1)(b) of the Conditions which states that in the event of ambiguities or discrepancies between the employer's requirements and the contractor's submission, the employer's requirements shall prevail. It can also be deduced from the form of tender where the contractor offers to design, construct and complete the works in conformity with the employer's requirements; and from clause 6(2)(a) which requires the contractor to submit designs and drawings showing that the works will comply with the employer's requirements.

The legal effect of this is to stop the contractor's submission being treated as a counter offer and itself forming the basis of the contractor's obligations under the contract.

Desirable as this may appear, it does have possible drawbacks. It would seem that under the Conditions the employer's requirements provide not only the basis of the contractor's obligations but that they also act as a limit on his obligations. That is to say, if the contractor's submission offers a better scheme or specification than that necessary to comply with the employer's requirements, the contractor may not be strictly bound by his scheme. It may be that in such circumstances the contractor enters the contract with a choice; to provide his own scheme or to provide the employer's scheme.

Against this it will no doubt be said that the contractor's submission forms part of the contract and the contractor's obligation is to construct the works in accordance with the contract. However, consider a very simple example. The employer's requirements specify softwood fencing; the contractor's submission offers hardwood fencing. There is then a discrepancy. According to clause 5(1)(b), the employer's requirements prevail.

Perhaps there will be cases where there is no discrepancy argument to save the contractor from the generosity of his submission but the message to employers is that they should be aware when evaluating tenders, that they may not be entitled to receive what appears to be on offer.

One way to overcome the problem is for the employer to upgrade his requirements to match the contractor's submission before the award of contract. This seems to be contemplated in the definition of the employer's requirements in clause 1(1)(e) which states, 'the requirements which are identified as such at the date of the award of the Contract'.

2.3 *Content of the employer's requirements*

Unless the D and C Conditions are supported by special conditions of contract or the like which are properly integrated into the contract, the employer's requirements will have to cover not only technical data but also contractual and administrative requirements which are either left open or not fully detailed in the Conditions. For example, details of any quality assurance system required; whether or not a contract price analysis is required; how interim payments are to be made.

The employer's requirements, therefore, are far more than a specification of work. They are the requirements not only for the work itself, but also for how it is to be carried out and how it is to be paid for.

Matters for inclusion

Matters which typically need to be considered for inclusion in the employer's requirements are:

- a description of the works
- the purposes of the works
- the location of the site with details of restrictions and access
- phasing requirements
- the involvement of other contractors
- details of site accommodation requirements
- information on statutory undertakers etc
- planning permissions and other statutory requirements
- public participation and public relations
- drawings and details
- technical specifications
- safety matters
- performance criteria
- performance testing
- rules on the format of the contractor's submission
- manuals and maintenance instructions
- contingencies and prime cost items
- requirements for a contract price analysis

- details of the scheme for interim payments
- schedules of rates for the valuation of variations
- professional indemnity insurances
- duty of care letters/collateral warranties.

Use of schedules

To facilitate the task of compiling the employer's requirements and to ease the burden of assimilating their contents it is good practice to use schedules or similar groupings. For example:

- the works
- the site and restrictions
- statutory matters
- drawings and specifications
- construction
- contingencies and prime cost sums
- payments and valuations
- insurances, bonds and guarantees.

Some standard forms, such as the IChemE Red and Green Books go a little further and require all details specific to the contract to be given in schedules; including such matters as completion times and liquidated damages.

In the D and C Conditions this is not necessary, and probably not desirable, because parts 1 and 2 of the appendix to the form of tender should cover much of the contractual detail specific to the contract.

2.4 *Detail of the employer's requirements*

Opinions differ as to whether design and construct contracting is best used when the employer specifies his requirements in fine detail or when he provides only broad outlines.

The argument against the fine detail approach is that it inhibits the contractor's freedom of choice and stifles innovation. But for many projects, and this is particularly true of civil engineering works, the employer often has to progress the design to a fairly advanced stage to obtain the consents and approvals necessary to proceed to tender stage. And then there is usually no going back on the employer's design and details.

The argument against the broad outline approach is that in a competitive environment it encourages poor quality in design and standards. There may be some truth in this in certain sectors of the construction industry but it should not apply to civil engineering with much effect.

One point in favour of the broad outline approach as far as the D and C Conditions are concerned is that there is less likelihood of discrepancy between the contractor's submission and the employer's requirements and the contractor will have less excuse for not providing what he has promised in his submission.

Typical specification details

As an example of the details which might be included in the employer's requirements by way of specification the following list gives the headings from the technical section of the employer's requirements on a highway design and construct contract:

- orders for land acquisition
- general road layout
- facilities for employer's representative
- fencing specification
- safety fence specification
- drainage specification
- extent of site clearance
- earthworks restrictions
- pavement specification
- traffic signs and road markings
- restrictions on roadworks
- testing
- landfill
- computer data
- access
- landscaping
- wildlife
- noise control
- statutory undertakers
- safety audits
- taking over procedures.

2.5 Omissions and economies

Employers should be aware that the D and C Conditions, particularly when used with a lump sum price, are not conducive to post-award financial savings.

Clause 51 of the Conditions does refer to omission variations but clause 52 on the valuation of variations has nothing to say on the valuation of omissions. This is discussed further in Chapter 16.

It appears that the most the employer can do if he wants to scale down his requirements after award or make economies of specification with a view to cost savings is to negotiate through his representative for the best deal the contractor will allow. Conciliation and arbitration are available as longstop methods of resolving disputes but they will normally be too late to assist the employer in making decisions on whether to proceed with the works in accordance with the original requirements or whether to strike a bargain with the contractor on a reduced price for a smaller or inferior scheme.

Chapter 3

The contractor's submission

3.1 *Introduction*

In traditional contracting, tenderers compete on the basis of price and little else. It may be true that the third lowest price is generally the right price but few contractors would welcome awards on that basis. Nor would employers, for when an experimental scheme was tried in Italy it led to a massive increase in prices as tenderers sought to avoid being the lowest.

In design and construct contracting, however, tenderers may feel with some justification that price is not everything and that the quality of their proposals should be taken into account. And generally, of course, quality is considered even amongst bids which all conform with the employer's requirements.

Price or quality

The problem for tenderers is to assess which of the two factors, price or quality, is of greatest importance to the employer. There is probably a presumption in favour of price unless the employer has a history of going for quality. And in the public sector, with strict audit rules, price usually takes preference over quality providing the employer's requirements are met in full.

So, recognising this, tenderers will generally aim to submit the lowest conforming price whilst at the same time putting forward a submission which indicates the best quality. There is an art in this which goes well beyond the glossy cover of the submission document. It has a lot to do with conveying confidence and demonstrating an understanding of the employer's requirements.

3.2 *Responding to opportunities*

The strength of a contractor's business can be gauged from the level of invitations to tender. As a rule of thumb, turnover will be in the order of 20% of the value of tenders submitted. Growth depends upon expanding the number of opportunities to tender; and this in turn depends upon the quality of marketing.

Contractors therefore are constantly searching for opportunities and there is a natural reluctance, particularly in difficult trading times, to turn anything down and to send enquiries back. The 'we can do that' philosophy is the marketing man's tool of trade. But in design and construct contracting this can be a mistake.

No contractor should take on the task of tendering for a design and construct contract without first considering:

- the risks of the project
- the capability of the organisation to undertake the project
- the contractual conditions imposed
- the timescale for submission of tenders
- the quality of the competition and the number of tenderers
- the costs of tendering
- the staffing resources available for preparing the bid
- the information to be obtained prior to tender
- the information required in the submission.

Unless analysis of the above factors is favourable, the contractor proceeds at his peril. The maxim used in traditional civil engineering contracting that every contract won is an opportunity for profit simply does not apply to design and construct contracting. There is too much transference of risk to the contractor and too little scope left for claims.

3.3 *The costs of tendering*

For contractors, tendering for a design and construct contract is an expensive business and it is not something to be embarked upon lightly.

Figures vary, but one national contractor has stated that whereas his average tender costs for traditional contracting are in the region of ¼% of the tender sum, his tender costs for design and construct are somewhere between 1% and 2% of the tender sum depending upon how much preparation has been put into the scheme by the employer and how much has been left by way of site investigation, design, planning, etc.

Number of tenderers

Not surprisingly with such high tender costs, most contractors will want to know before they commit themselves to tendering how many tenderers are on the list. And any number greater than four is usually regarded as unacceptable for design and construct contracting.

The economics of this are simple enough. With a one in four chance of winning a contract and average tender costs at the middle range at 1½%, then the contractor needs to earn a 6% margin from the work he wins just to cover his organisation's tendering overheads. This margin is scarcely obtainable at the best of times in some markets but clearly with six or eight tenderers, as would be commonplace in traditional contracting, the required margin would reach impossible proportions.

Cost to the employer

There is in these numbers something of a warning to employers. Namely that in the long term everything has to be paid for. Contractor's design is not a free service and the costs have to be recovered in tender margins. And so whilst an employer might appear to get the benefit of choice of say four alternative designs, whilst paying only for one, this is an illusion. All four designs have to be paid for; and the employer pays through higher margins in design and construct contracting than in traditional contracting.

Sharing the costs

Contractors have only a limited number of opportunities to share tendering costs if they are to protect the confidentiality of their own submissions. But in two areas cost sharing is frequently attempted.

Firstly, there is the prospect of sharing costs with professional practices; the consultants, quantity surveyors etc involved in the preparation of the submission. The question is how far they are prepared to work on a no win/no fee (no hay/no pay) basis.

Then there is the prospect of sharing site investigation costs with other tenderers. One device is for the tenderers to commission a single site investigation report which is made available to all with the cost divided equally. Each tenderer then includes in his bid price the full cost of the site investigation and reimburses the unsuccessful tenderers their shares on award of the contract.

3.4 The management of tendering

Tendering for design and construct contracts is a long way removed from the routine of estimating for traditional contracts. In assembling the contractor's submission the estimator is no more than a cog in the wheel and the contractor normally needs to look outside his estimating department for the management of his submission.

Submission strategy

Decisions at senior levels are required to devise the strategy for each submission in line with the objectives of the company and the characteristics of the particular tender.

Matters to consider are:

- appointment of the submission manager
- selection of the design team
- the involvement of external specialists
- the method of pricing
- the tender programme
- the tender budget

The submission manager

The submission manager is given various titles of bid manager, proposals manager, and the like. Perhaps, submission co-ordinator comes closest to describing his function.

His role is to liaise between the designers, the specialist sub-contractors, the estimators and the planners and to co-ordinate their activities. He has to ensure that all external constraints are investigated and that the tender includes for all the contractor's obligations.

Not least the submission manager has to maintain a good relationship with, and the confidence of, the potential client. Many a design and construct tender has been lost because the contractor's inadequacies for the contract in mind have been revealed in discussions and meetings with the contractor's staff before the tender has been submitted. It is a good rule for only the submission manager to have contact with the client at this stage.

Selection of the design team

Only a few civil engineering contractors have sufficient in-house resources to undertake design. For the majority, external consultants need to be appointed early in the tendering process.

Selection of the right consultants is fundamental to the contractor's success – both in tendering and in carrying out any contract awarded. Success depends upon having a commercial edge and that has to start with the design.

So the first criteria in the selection process is likely to be, how commercially minded is the consultant. Can he produce competitive and buildable designs and perform to programme? Other criteria will include:

- previous relationships
- previous experience in design and construct
- previous experience of the type of work
- standing in the eyes of the client
- locally based resources (area office)
- fee levels
- insurance cover
- willingness to sign warranties.

Alternative proposals

Tenderers are sometimes invited to submit alternative proposals. But even when not so invited it is inevitable that the design process will produce ideas from time to time which do not conform with the employer's requirements but seem to offer a better solution. The problem is to get this alternative considered.

Contractors should not be discouraged by tendering rules which state that all non–conforming bids will be rejected. The proper course it to submit both a conforming bid and the alternative proposals. This enables the adjudicating panel to make comparisons on a like for like basis whilst leaving open the door for further discussion on the alternative proposals.

3.5 Content of the contractor's submission

The form and content of the contractor's submission will vary from project to project but as a minimum it is likely to contain:

- design proposals
  - names of the designers
  - a written statement of design philosophy
  - drawings and details
  - specifications
  - details of specialist products and sub-contractors
- construction proposals
  - programmes and networks
  - method statements
  - arrangements for supervision/QA etc.
  - safety policies
- financial details
  - the contract price
  - the contract price analysis
  - schedules of daywork rates
- management structure
  - the organisation chart
  - the names and CVs of key personnel
- details of any assumptions made in tendering or qualifications made in the submission.

Additionally many contractors bind into the submission document information on the company's background and details of similar contracts undertaken. Although its worth cannot be measured a glossy cover is considered essential.

3.6 The contractor's submission in the Design and Construct Conditions

Definition

Clause 1(1)(f) of the D and C Conditions defines the contractor's submission as:

- the tender and
- all documents forming part of the offer
- together with such modifications and additions

- as may be agreed between the parties prior to the award of the contract.

Two points of considerable importance emerge from this definition. One that negotiations between the submission of the tender and the award of the contract are not only contemplated by, but are provided for in the Conditions. The other is that the contractor's submission is fixed at the date of award of the contract and, unlike the employer's requirements, the defined term does not include for subsequent variations.

Pre-award modifications and additions

The D and C Conditions are not intended for formal two stage tendering. That is the process where tenderers submit a firm price for the design contract and the construction contract is subsequently negotiated. Clearly, the D and C Conditions apply to single stage tendering where the price given covers both design and construction.

However, even in single stage tendering there are usually points to be clarified and documents to be finalised before a contract is awarded. Such matters may be purely technical but they may also be financial, particularly if the contractor has included provisional sums, prime cost sums or contingencies in his price.

The recording of any modifications or additions to the contractor's submission needs to be undertaken with caution and precision by both parties. There will be nothing in the form of tender to cover these matters (which are post-tender) and although they can be listed in the form of agreement, that document is only likely to be signed when the contract is to be executed as a deed. In any event, signing of the agreement usually follows the award of the contract as a formality so it confirms rather than defines, the terms of the contract.

Variations to the contractor's submission

The definition in clause 1(1)(f) of the Conditions does not allow for any variation in the contractor's submission. Nor does clause 51 of the Conditions which deals only with variations to the employer's requirements. This approach is in marked contrast to most other design and construct forms of contract which recognise that the contractor may wish to change, or may be obliged for practical measures to implement a change, to the details in his submission.

There appears by this to be an unnecessary and potentially unworkable rigidity in the D and C Conditions. The point is discussed in detail in later chapters.

Status of the contractor's submission

Apart from in clause 1 (definitions) and clause 5 (contract documents) the Conditions do not refer anywhere to the contractor's submission. This by itself is a fair indication that the contractor's submission does not occupy of its own accord a position of much status in the Conditions.

The position with the employer's requirements however is quite different. Although taken together the contractor's submission and the employer's requirements comprise the contract, it is the employer's requirements that are referred to repeatedly throughout the Conditions and that clearly form the true basis of the contractor's obligations. The contractor's submission is imported into the contractor's obligations only by reference to 'the Contract'.

Moreover, as discussed in Chapter 2, clause 5(1)(b) of the Conditions states, for the avoidance of any doubt, that in the event of any ambiguities or discrepancies between the employer's requirements and the contractor's submission, it is the employer's requirements which shall prevail.

Ambiguities

Clause 5(1)(d) of the Conditions states that any ambiguities or discrepancies within the contractor's submission shall be resolved at the contractor's expense.

This quite rightly makes the contractor responsible for the accuracy of his submission but it does not go as far as the corresponding provision in the standard design and build form (JCT 81) by stating what is meant by 'resolved'.

In JCT 81 if there are discrepancies within the contractor's proposals, the employer can choose between the discrepant items at no additional cost. This is not obviously the position under the D and C Conditions and it is doubtful if it can be implied. It seems more likely that if the contractor's submission contains ambiguities or discrepancies but they do not breach the employer's requirements, then it is the contractor who decides. This comes back to the point that the employer may not always get what he thinks was on offer in the contractor's submission.

Chapter 4

Obligations and responsibilities for design

4.1 *Introduction*

Much of the comment in journals and at seminars on the D and C Conditions has focused, quite naturally, on the implications for the contractor in taking on responsibility for design. And the topic which in particular has dominated debate is whether the contractor's responsibility is on a skill and care basis or a fitness for purpose basis.

Skill and care/fitness for purpose

The point of this debate is that if the contractor's responsibility for design is limited to skill and care corresponding to that of a professional designer, then negligence must be proved to establish breach of duty – whether in contract or in tort. That is design failure may occur but the contractor will not be liable provided he has used proper skill and care. However, if the contractor's responsibility is fitness for purpose that is a strict responsibility. The contractor may have used all proper skill and care but if the specified contractual objective is not achieved the contractor will be liable to the employer for damages.

Generally, therefore, skill and care is a lower standard of responsibility than fitness for purpose. Consequently contracts requiring fitness for purpose are thought to be more onerous on the contractor than skill and care contracts. Additionally they create a problem for the contractor in that it is difficult for him to pass on responsibilities of fitness for purpose to professional designers.

The D and C Conditions appear to strive towards skill and care but probably fall some way short of achieving the objective. If any consensus can be drawn from the views of lawyers on the matter it is probably that the Conditions leave plenty of scope for dispute.

Limited effect of fitness for purpose

Essential as it is that the contractor's legal liabilities for design are fully understood, the practical importance of these liabilities needs to be kept in perspective. Only in exceptional cases of collapse or failure will the question of skill and care responsibility or fitness for purpose arise. In most cases, evidence of negligence in design will catch the contractor on the lower standard of skill and care and argument on whether the higher standard of fitness for purpose applies will not be relevant.

So true is this that even in one of the leading legal cases on contractor's design responsibility – *IBA* v. *EMI Electronics Ltd* (1980) – the observations of the House of Lords on fitness for purpose were not strictly necessary since the House upheld the findings of negligence by the judge at first instance. In that case, of which more will be said later in this chapter, a 1250 foot high television mast at Emley Moor in Yorkshire collapsed within four months of completion. It was the highest mast of its kind in the world and of novel cylindrical design. The defence to the charge of negligence – that the design was beyond the frontiers of professional knowledge at that time – was rejected. The designers were held to be negligent in not having considered the effect of asymmetric ice loading on the stays and vortex shedding.

Problems in the Design and Construct Conditions

Too much attention can, therefore, be given to the fitness for purpose point and too little perhaps to other aspects of design responsibility in the D and C Conditions. And there are matters which will arise in the ordinary use of the Conditions which will create difficult practical and legal problems.

This seems inevitable from the requirement that the contractor should accept responsibility for the employer's design; from the requirement that the contractor should obtain the consent of the employer's representative to his designs and drawings; from the powers of the employer's representative in respect of other approvals; and from the absence of any express entitlement allowing the contractor to vary his design when he considers it necessary. The Conditions are silent on how disagreement between the parties on such matters is to be resolved. Conciliation or arbitration may be suggested but these are really processes designed to deal with the financial consequences of disputes. They are not the ideal vehicles for making decisions on how the works should proceed.

Comparison with other standard design and construct forms

The following table shows for comparison purposes how the contractor's obligations and responsibilities for design are treated in various standard forms of contract used in the construction industry.

Standard form	Extent of the contractor's design obligation	Basis of the contractor's design responsibility
ICE D and C Conditions	the whole design (but see later sections)	skill and care (but see later sections)
JCT 81	to complete the design	skill and care
ACA/BFP	to complete the design	fitness for purpose for the parts designed
New Engineering Contract	to design the parts the contract requires the contractor to design	fitness for purpose with optional limitation to skill and care
The Department of Transport Design and Construct form	the whole design	fitness for purpose
GC/Works/1 (edition 3) Single Stage Design and Build	the whole design	skill and care
IChemE Red Book	to complete the design	fitness for purpose
IEE/IMechE MF/1	the whole design unless otherwise disclaimed	fitness for purpose with limiting provisions

4.2 *Terminology*

Not all the key terms used in design and construct contracts are fixed by definition or have precise legal meaning. Even in dictionaries some terms are expressed only in generalities.

For understanding of the D and C Conditions, four terms deserve brief comment:

- design
- obligation
- responsibility
- liability.

Meaning of design

Amongst the many meanings of design in the *Shorter Oxford Dictionary* the ones most applicable to design in construction are:

- the preliminary conception of an idea that is to be carried into effect by action
- to make the preliminary sketch of; to make the plans and drawings necessary for the construction of.

Design, in short, is the accumulation of ideas and details which go into the production of an artefact or the construction of a project.

Thus in a design and construct project, design can rarely be confined to the input of the contractor. The employer's requirements, by providing plans, details, specifications and the like, will also be part of the design. Only if the employer avoids describing the finished works by limiting his requirements to performance criteria can the contractor have complete control of design. And even in process and plant contracts where performance is often all that matters it is difficult for employers (or purchasers as they are known in such contracts) to say nothing about the appearance or style of the works.

Obligations, responsibilities and liabilities

Obligation is described by the *Shorter Oxford Dictionary* as (amongst other things):

> 'The action of binding oneself by oath, promise or contract to do or forbear something; a binding agreement; also that to which one binds oneself, a formal promise'.

And in *Osborne's Concise Law Dictionary* the definition of obligation is:

> 'A duty; the bond of legal necessity which binds two or more determinate individuals. It is limited to legal duties arising out of a special personal relationship existing between them, whether by reason of a contract or a tort, or otherwise'.

Responsibility is described in the *Shorter Oxford Dictionary* as 'the state or fact of being responsible'. Responsible is described as 'answerable, accountable, liable to be called to account'.

Liability, according to *Osborne's Concise Law Dictionary* is 'subjection to a legal obligation; or the obligation itself. He who commits a wrong or breaks a contract or trust is said to be liable or responsible for it'.

Clearly in legal terms there is not much to distinguish an obligation from a responsibility and the words 'obligation' and 'responsibility' are often taken as synonymous. But in ordinary language there is a difference. An obligation is a burden to be undertaken; a responsibility is a burden to be carried.

Thus in the D and C Conditions, the contractor has an obligation to construct the works; but he carries responsibility for damage to the works. The employer has an obligation to give possession of the site; but he has responsibility for the defaults of his employees.

Obligations of the contractor for design

This distinction between obligations and responsibilities is drawn to show how the phrases 'the obligations of the contractor' and 'responsibility for design' apply in construction contracts. And the point of interest, in relation to design and construct contracts, is what is the contractor's obligation in respect of design and what is his responsibility?

The table above shows that in some design and construct contracts, including the D and C Conditions, the contractor's obligation is stated to be 'to design the works'. In others it is stated to be 'to complete the design of the works'.

It is difficult to see how, from the meaning of design considered above, the contractor can have anything other than an obligation to complete the design except in the circumstances where the employer does no more than state performance criteria. The contractor cannot undertake that which has already been undertaken. And in most design and construct contracts the employer will have undertaken design in formulating his requirements. In the D and C

Conditions, clause 8(2)(b) refers directly to parts of the works designed by, or on behalf of, the employer.

It is probable that contracts which state that the contractor has an obligation to design the works do so to emphasise that the contractor has responsibility for the whole of the design. Those contracts which state specifically that the contractor's obligation is to complete the design, usually, although MF/1 is an exception, limit the contractor's responsibility for design to the parts he has designed.

Design responsibility of the contractor

The point considered here is not whether the basis of the contractor's responsibility is skill and care or fitness for purpose, that is considered later, but the extent of the contractor's design responsibility.

Thus in the D and C Conditions it is possible to question whether the contractor can have an obligation to design the works (as opposed to an obligation to complete the design) but it is not possible to question that the Conditions intend the contractor to take responsibility for the whole of the design. It is expressly stated in clause 8(2)(b) that the contractor shall accept responsibility for any design included in the employer's requirements.

This provision is not unique to the D and C Conditions but it is certainly onerous on the contractor. And it represents the transference of risk from the employer to the contractor which could be potentially unfair. Thus if the employer has been negligent in his design the responsibility appears to fall on the contractor. However the transfer of responsibility is only between the employer and the contractor and it does not affect the employer's responsibilities to third parties or under statute.

But unfair or not, if contractors agree to accept responsibility for the employer's design, that is what they do and not too much reliance should be placed on the Unfair Contract Terms Act 1977 to save them from the consequences of their actions. The Act applies only to limitations on liability for negligence and not to the transference of liability.

4.3 General principles of design liability

Any designer, whether he be a professional designer or a contractor designer, can have liabilities in contract or in the tort of negligence.

The doctrine of privity of contract confines liabilities in contract to the parties to the contract; but in tort liabilities can extend to third parties.

Liability in tort

It is not intended in this book to deal with liability in tort other than briefly. The key principles are as follows:

- there must be a duty of care
- there must be breach of that duty
- the breach must result in damage or injury.

The modern law of the tort of negligence derives from the famous snail in the ginger beer bottle case – *Donoghue* v. *Stevenson* (1932). In that case Lord Aitkin said:

> 'The rule that you are to love your neighbour becomes in law, you must not injure your neighbour; and the lawyer's question, Who is my neighbour? receives a restrictive reply. You must take reasonable care to avoid acts or omissions which you can reasonably foresee would be likely to injure your neighbour. Who, then, in law is my neighbour? The answer seems to be – persons who are so closely and directly affected by my act that I ought reasonably to have them in contemplation as being so affected when I am directing my mind to the acts or omissions which are called in question.'

For half a century the law developed adding new categories of negligence with physical damage and economic loss joining injury as grounds for action. But in the 1980s the courts began a reversal of the trend, culminating in the case of *Murphy* v. *Brentwood District Council* (1990). The position now is that for a physical damage claim in negligence the physical damage must be damage to an article or property other than the article or property to which the negligence itself applies – *D & F Estates* v. *Church Commissioners for England* (1988). And, except in cases of special proximity such as *Hedley Byrne* v. *Heller* (1964), pure economic loss is no longer recoverable in negligence – *Murphy* v. *Brentwood* (1990)

The question of whether, when there is a duty in contract, there is a parallel duty in tort is not decided with certainty. At one time it seemed from the case of *Bagot* v. *Stevens Scanlan & Co.* (1966) that the duties owed by professional men arose only in contract and not in tort but that position was disturbed by the findings in *Esso Petroleum Co. Ltd* v. *Mardon* (1976) and *Batty* v. *Metropolitan Property Realisation Ltd* (1978). Then in 1986 the Privy Council in *Tai Hing Cotton Mill Ltd* v. *Liu Chong Hing Bank Ltd* appeared to revert to *Bagot* when Lord Scarman said:

> 'Their Lordships do not believe that there is anything to the advantage of the law's development in searching for a liability in tort where the parties are in a contractual relationship. This is particularly so in a commercial relationship.'

However the recent case of *Barclays Bank plc* v. *Fairclough Building Ltd* (1993) shows that the *Tai Hing* ruling may be of only limited application. In the *Barclays Bank* case high speed water jetting to clean an asbestos sheeted roof, led to asbestos entering the Bank's premises. The judge held that the Bank did have a claim in tort. There was no defect in the contract work (i.e. cleaning the roof); what was complained of was damage to other property.

Negligence

The meaning of negligence was eloquently stated in the case of *Bolam* v. *Friern Hospital Management Committee* (1957). Bolam, a patient in a mental hospital, was given electro-convulsive therapy without any restraint or sedation and as a result was injured. Mr Justice McNair said:

> 'I must tell you what in law we mean by "negligence". In the ordinary case which does not involve any special skill, negligence in law means a failure to do some act which a reasonable man in the circumstances would do, or the doing of some act which a reasonable man in the circumstances would not do; and if that failure or the doing of that act results in injury, then there is a cause of action. How do you test whether this act or failure is negligent? In an ordinary case it is generally said you judge it by the action of the man in the street. He is the ordinary man. In one case it has been said you judge it by the conduct of the man on the top of a Clapham omnibus. He is the ordinary man. But where you get a situation which involves the use of some special skill or competence then the test as to whether there has been negligence or not is not the test of the man on the top of the

Clapham omnibus, because he has not got this special skill. The test is the standard of the ordinary skilled man exercising and professing to have that special skill. A man need not possess the highest expert skill; it is well established law that it is sufficient if he exercises the ordinary skill of an ordinary competent man exercising that particular art'.

Skill and care

The duty of a professional man to exercise skill and care was established long before the *Bolam* case. In *Lanphier* v. *Phipos* (1838) it was said: 'Every person who enters in to a learned profession undertakes to bring to the exercise of it a reasonable degree of care and skill'.

What the *Bolam* case illustrated was the level of skill required to avoid negligence – the skill of any ordinary practitioner. For the construction industry no better exposition will be found than that of Lord Justice Bingham in the case of *Eckersley* v. *Binnie & Partners* (1988). He said:

> 'The law requires of a professional man that he live up in practice to the standard of the ordinary skilled man exercising and professing to have the special professional skills. He need not possess the highest expert skill; it is enough if he exercises the ordinary skill of an ordinary competent man exercising his particular art. So much is established by *Bolam* v. *Friern Hospital Management Committee* [1957] 1 WLR 582 which has been applied and approved time without number. "No matter what profession it may be, the common law does not impose on those who practise it any liability for damage resulting from what in the result turn out to have been errors of judgment, unless the error was such as no reasonable well-informed and competent member of that profession could have made" (*Saif Ali* v. *Sydney Mitchell & Co.* [1980] AC 198 220D, per Lord Diplock).
>
> From these general statements it follows that a professional man should command the corpus of knowledge which forms part of the professional equipment of the ordinary member of his profession. He should not lag behind other ordinarily assiduous and intelligent members of his profession in knowledge of new advances, discoveries and developments in his field. He should have such awareness as an ordinarily competent practitioner would have of the deficiencies in his knowledge and the

limitations on his skill. He should be alert to the hazards and risks inherent in any professional task he undertakes to the extent that other ordinarily competent members of his profession would be alert. He must bring to any professional task he undertakes no less expertise, skill and care than any other ordinarily competent members of is profession would bring, but need bring no more. The standard is that of the reasonable average. The law does not require of a professional man that he be a paragon, combining the qualities of polymath and prophet.

In deciding whether a professional man has fallen short of the standards observed by ordinarily skilled and competent members of his profession, it is the standards prevailing at the time of acts or omissions which provide the relevant yardstick. He is not ... to be judged by the wisdom of hindsight. This of course means that knowledge of an event which happened later should not be applied when judging acts and/or omissions which took place before that event ...;... it is necessary, if the defendant's conduct is to be fairly judged, that the making of [any] retrospective assessment should not of itself have the effect of magnifying the significance of the ... risk as it appeared or should reasonably have appeared to an ordinarily competent practical man with a job to do at the time.'

A higher duty of skill and care

The proposition that a client who deliberately obtains and pays for someone with specially high skills is entitled to expect a higher level of skill than that which is ordinary was considered in the unusual case of *Wimpey Construction Ltd* v. *Poole* (1984) of which more is said in Chapter 11. Wimpey, in an attempt to claim off their insurers for failure of an in-house designed project, sought to argue their own negligence by claiming that a company of their standing should be judged by more stringent and exacting standards than an ordinary designer. The judge, however, could find no precedent to accept such a proposition.

A second point of interest in the *Wimpey* case was the suggestion that it is the duty of a professional man to exercise reasonable care in the light of his actual knowledge and that the question whether he exercised reasonable care cannot be answered by reference to a lesser degree of knowledge than he had, on the grounds that the ordinarily competent practitioner would only have had that lesser degree of knowledge. The judge accepted this but pointed out that where a professional man has knowledge, and acts or fails to act in a way which, having that knowledge, he ought reasonably to foresee would cause damage, then, if the other aspects of duty are present, he would be liable in negligence by virtue of the direct application of the test in *Donoghue* v. *Stevenson*.

And, summing-up the test of negligence, the judge in the *Wimpey* case said this:

> 'Lastly, while considering the test of negligence, I should advert to the necessity of avoiding hindsight and of judging the conduct of the designer or designers of the quay wall by the standards of the time at which it was designed, not by any later standards, still less the standards of today, and of disregarding the fact that we now know the consequences of the design.'

State-of-the-art defence

The policy of the courts in avoiding hindsight when judging negligence and applying the test of the professional knowledge of the time the design was carried out leads to what is known as the state-of-the-art defence. That is the designer applied the knowledge of the time.

This defence is open to challenge on a number of grounds; one of which is that there may be a continuing duty on a designer to check his design and to revise in the light of new knowledge. How far this extends to a duty to warn after completion is a subject beyond the scope of this book since the law is still developing and definite principles are hard to define.

Another possible challenge is that the practice of the profession at the time of the design was itself incorrect. This challenge has been successfully mounted in some medical negligence cases, most notably, *Sidaway* v. *Governors of Bethlem Royal Hospital* (1985).

Innovation and special circumstances

'The law requires even pioneers to be prudent.' Not my words but those of Lord Edmund-Davies in the *IBA* v. *EMI* case.

And this is the third challenge to the state-of-the-art defence. It will not do to say, as was said in that case, that the work was both at and beyond the frontier of professional knowledge at that time.

This is how Lord Edmund-Davies explained the matter:

'What is embraced by the duty to exercise reasonable care must always depend on the circumstances of each case. They may call for particular precautions (*Redhead* v. *Midland Railway Co* (1869) LR4 QB, 379 at 393). The graver the foreseeable consequences of failure to take care, the greater the necessity for special circumspection (*Paris* v. *Stepney Borough Council* [1951] AC 367), and "Those who engage in operations inherently dangerous must take precautions which are not required of persons engaged in the ordinary routine of daily life" (*Glasgow Corporation* v. *Muir* [1943] AC 448, per Lord Macmillan at 456). The project may be alluring. But the risks of injury to those engaged in it, or to others or to both, may be so manifest and substantial, and their elimination may be so difficult to ensure with reasonable certainty that the only proper course is to abandon the project altogether. Learned counsel for BICC appeared to regard such a defeatist outcome as unthinkable. Yet circumstances can and have at times arisen in which it is plain common sense, and any other decision foolhardy. The law requires even pioneers to be prudent.'

Liability in contract

Contractual liability for design may arise from:

- the express terms of the contract
- terms implied by common law
- terms implied by statute.

This applies to both design and construct contracts and the contracts of engagement of professional designers.

The express terms of the D and C Conditions and some of the other more often used design and construct contracts are considered later in this chapter. Brief comment is also made on the position of professional designers under the heading of collateral warranties.

Terms implied by common law

The conditions which must be satisfied for a term to be implied into a contract on common law principles were set out by Lord Simon in the case of *BP Refinery Ltd* v. *Shire of Hastings* (1977). They are:

- it must be reasonable and equitable
- it must be necessary to give business efficacy to the contract, so no term will be implied if the contract is effective without it
- it must be so obvious that 'it goes without saying'
- it must be capable of clear expression
- it must not contradict any express term of the contract.

Important implied terms relating to construction contracts are:

- In a contract for labour –
  to do work in a good and workmanlike manner. *Duncan* v. *Blundell* (1820) where it was said:

  > 'Where a person is employed in a work of skill, the employer buys both his labour and his judgment; he ought not to undertake the work if it cannot succeed, and he should know whether it will or not. Of course it is otherwise if the party employing him choose to supersede the workman's judgment by using his own.'

  Approved in *Young & Marten* v. *McManus Childs* (1969).

- In a contract for labour and materials –
  to supply materials of good quality and reasonably fit for their purpose. *Myers* v. *Brent Cross Service Co.* (1934) where it was said:

  > 'A person contracting to do work and supply materials warrants that the materials which he uses will be of good quality and reasonably fit for the purpose for which he is using them, unless the circumstances of the contract are such as to exclude any such warranty.'

  Approved in *Young & Marten* v. *McManus Childs* (1969).

- In a contract for a house –
  that it should be reasonably fit for human habitation *Hancock* v. *Brazier* (1966).

- In a contract for design and construction –
  that the finished works will be reasonably fit for their intended purpose *IBA* v. *EMI* (1980).

Terms implied by statute

The statutes with particular relevance to construction contracts and the subject of design liability are (amongst others):

- The Sale of Goods Act 1979
- The Supply of Goods and Services Act 1982.

Also of some relevance is the Unfair Contract Terms Act 1977 which limits the freedom of suppliers to contract out of statutory obligations.

The Sale of Goods Act 1979 imposes into contracts of sale various warranties, including:

- goods supplied shall correspond with their description
- goods supplied shall be of merchantable quality
- goods supplied shall be reasonably fit for any purpose made known to the supplier where the purchaser relies on the skill and care of the supplier.

The Supply of Goods and Services Act 1982 applies to contracts which are not solely for the supply of goods but which include services (such as design). The Act provides amongst other things for the following terms to be implied into contracts:

- that goods will correspond with description
- that goods will be of merchantable quality
- that a supplier acting in the course of business will carry out the service with reasonable skill and care
- that nothing in the Act prejudices any rule of law which imposes a stricter duty than implied by the Act.

4.4 *Fitness for purpose*

> 'The virtue of an implied term of fitness for purpose is that it prescribes a relatively simple and certain standard of liability based on the 'reasonable' fitness of the finished product, irrespective of considerations of fault and of whether its unfitness derived from the quality of work or materials or design'.

Judge John Davies in *Viking Grain Storage* v. *T H White Installations Ltd* (1985).

Implied terms

The law does not normally imply terms of fitness for purpose into contracts for the employment of professional men. This is how Lord Denning MR explained the position in *Greaves & Co. (Contractors) Ltd* v. *Baynham Meikle & Partners* (1975):

> 'Apply this to the employment of a professional man. The law does not usually imply a warranty that he will achieve the desired result, but only a term that he will use reasonable care and skill. The surgeon does not warrant that he will cure the patient. Nor does the solicitor warrant that he will win the case. But, when a dentist agrees to make a set of false teeth for a patient, there is an implied warranty that they will fit his gums'.

The position of a contractor, however, is significantly different. The law does imply terms of fitness for purpose into contracts where the contractor supplies materials and the employer relies on the contractor's judgment in selection – *Myers* v. *Brent Cross Service Co.* (1934). And the law implies terms in contracts to build houses that the houses will be fit for habitation – *Miller* v. *Cannon Hill Estates* (1931).

Fitness for purpose in design and construct contracts

From the starting position above the question is – does the law imply a term of fitness for purpose into every design and construct contract? Well clearly it will not do so if there is an express term opposing such an implication but otherwise there is strong presumption in favour of the implied term. The following extracts from legal judgments illustrate this.

Lord Denning MR in *Greaves (Contractors) Ltd* v. *Baynham Meikle & Partners* (1975).

> 'This case arises out of a new kind of building contract called a "package deal". The building owners were Alexander Duckham & Co. Ltd. They wanted a new factory, warehouse and offices to be constructed at Aldridge in Staffordshire. The warehouse was needed as a store in which barrels of oil could be kept, until they were needed and despatched, and in which they could be moved safely from one point to another. The "package deal" meant that the building owners did not employ their own architects or

engineers. They employed one firm of contractors to do everything for them, called Greaves & Co (Contractors) Ltd. It was the task of the contractors, not only to provide the labour and materials in the usual way, but also to employ the architects and engineers as sub-contractors. The contractors were to do everything as a "package deal" for Alexander Duckham & Co.

Now, as between the building owners and the contractors, it is plain that the owners made known to the contractors the purpose for which the building was required, so as to show that they relied on the contractor's skill and judgment. It was, therefore, the duty of the contractors to see that the finished work was reasonably fit for the purpose for which they knew it was required. It was not merely an obligation to use reasonable care. The contractors were obliged to ensure that the finished work was reasonably fit for the purpose. That appears from the recent cases in which a man employs a contractor to build a house – *Miller* v. *Cannon Hill Estates Ltd* [1931] 2KB 113; *Hancock* v. *B W Brazier (Anerley) Ltd* [1966] 1 WLR 1317. It is a term implied by law that the builder will do his work in a good and workmanlike manner; that he will supply good and proper materials; and it will be reasonably fit for human habitation. Similarly in this case Greaves undertook an obligation towards Duckham that the warehouse should be reasonably fit for the purpose for which, they knew, it was required, that is, as a store in which to keep and move barrels of oil.'

Lord Scarman in *IBA* v. *EMI* (1980).

'In the absence of a clear, contractual indication to the contrary, I see no reason why one who in the course of his business contracts to design, supply and erect a television aerial mast is not under an obligation to ensure that it is reasonably fit for the purpose for which he knows it is intended to be used. The Court of Appeal held that this was the contractual obligation in this case, and I agree with them. The critical question of fact is whether he for whom the mast was designed relied upon the skill of his supplier (i.e. his or his sub-contractor's skill) to design and supply a mast fit for the known purpose for which it was required.

Counsel for the appellants, however, submitted that, where a design, as in this case, requires the exercise of professional skill, the obligation is no more than to exercise the care and skill of the ordinarily competent member of the profession. Although it might be negligence today for a constructional engineer not to realise the danger to a cylindrical mast of the combined forces of vortex shedding and asymmetric ice loading of the stays, he submitted that it could not have been negligence before the collapse of this mast: for the danger was not then appreciated by the profession. For the purpose of the argument, I will assume (contrary to my view) that there was no negligence in the design of the mast, in that the profession was at that time unaware of the danger. However, I do not accept that the design obligation of the supplier of an article is to be equated with the obligation of a professional man in the practice of his profession. In *Samuels* v. *Davis* [1943] KB 526, the Court of Appeal held that, where a dentist undertakes for reward to make a denture for a patient, it is an implied term of the contract that the denture will be reasonably fit for its intended purpose. I would quote two passages from the judgment of du Parcq LJ. At p. 529 he said (omitting immaterial words):

> "...if someone goes to a professional man ... and says: 'Will you make me something which will fit a particular part of my body?' ... and the professional gentleman says: 'Yes', without qualification, he is then warranting that when he has made the article it will fit the part of the body in question."

And at p. 530 he added:

> "If a dentist takes out a tooth or a surgeon removes an appendix, he is bound to take reasonable care and to show such skill as may be expected from a qualified practitioner. The case is entirely different where a chattel is ultimately to be delivered."

I believe the distinction drawn by du Parcq LJ to be a sound one. In the absence of any terms (express or to be implied) negativing the obligation, one who contracts to design an article for a purpose made known to him undertakes that the design is reasonably fit for the purpose.'

Judge John Davies in *Viking Grain Store* v. *T H White Installations Ltd* (1985).

'The suggestion that matters of design should be regarded as involving no higher duty than that of reasonable care was put forward and rejected in *IBA* v. *EMI* (1978) 11 BLR 38, where the judgment was delivered by Roskill LJ, where the Court of Appeal could see no good reason for importing into a contract of this nature a different obligation in relation to design from that which plainly exists in relation to materials. To find otherwise in this particular case, where Viking clearly relied, in all aspects, including design, on the skill and judgment of White to produce an end result would, in my view, be to destroy the whole basis of the bargain. The obligation to design a product fit for its purpose is already tempered by the fact that only "reasonable" fitness is demanded; to add to that a requirement of proof of lack of due care seems to me to emasculate, and magnify the uncertainty of, the obligation to such an extent as would be neither acceptable nor realistic in a commercial transaction.'

Can fitness for purpose be implied into professional contracts?

The *Greaves* case mentioned above led to some concern amongst professionals that fitness for purpose terms might be implied into contracts of engagement other than those where there was an element of supply.

The Court of Appeal in the *Greaves* case went out of its way to dispel this notion. Lord Justice Geoffrey Lane made these comments:

> 'At first sight [the judge] does appear to be saying that there is implied by law a higher form of duty than that set out in *Bolam's* case, a duty which is nevertheless lower than a warranty. If the learned judge indeed had said that, it would have been wrong; because there is no such duty. We are told by learned counsel that neither side at the trial advanced such a proposition, and I do not believe that the judge was intending to say that there was such a duty. What he was intending to convey was that there may be special circumstances in any particular case which require the reasonably careful expert to take special steps before his duty under *Bolam's* case can be said to have been discharged.'

and

> 'So far as the implied warranty is concerned it was suggested that to uphold the judge's finding of implied warranty in this case would mean that every professional man would be warranting the successful outcome of his endeavours. Nothing of the sort. All

the judge was in fact saying was that on the facts proved before him a warranty was to be implied that the building as designed by the defendants would be fit for the purpose which the plaintiffs stipulated; and one has only to read those passages in the evidence to which my Lord has already referred and also the defence as originally drafted to see that there was ample evidence upon which the judge could come to the conclusion that he did.

The suggestion that by reason of this finding every professional man or every consultant engineer by implication of law would be guaranteeing a satisfactory result is unfounded.'

The *Greaves* case was considered in *Hawkins* v. *Chrysler (UK) Ltd and Burne Associates* (1986) where an employee of Chrysler who slipped on a wet shower room floor sued for damages. Lord Justice Neill made these comments:

'I should say a few words, however, on the question of the implied warranty. It is clear from the decision of this court in *Greaves & Co (Contractors) Ltd* v. *Baynham Meikle & Partners* [1975] 1WLR 1095; 4 BLR 56 that in certain circumstances facts may establish the existence of an implied warranty by an independent consulting engineer that a building or structure designed by him will be fit for a stated purpose.

The responsibility undertaken by the consulting engineer in such a case may, it seems, extend to the suitability of materials specified by him for incorporation into his design.

It is also clear that a professional man may be liable on the basis of an implied warranty of fitness for purpose if his contract extends not merely to the design of an article, but also to its supply or manufacture. (See, for example, *Samuels* v. *Davis* [1943] KB 526.) Furthermore it is now established that a contractor, who has agreed to design and erect a building or other structure, may be liable for breach of an implied warranty that the building or structure is fit for a particular purpose, even though the contractor took no part in the design work, which was carried out by a specialist sub-contractor (see *IBA* v. *EMI* [1980] 14 BLR 1).

The question which arises in the instant case, however, is whether such a warranty of fitness for purpose is to be implied where there are no special facts; where there is no evidence that the professional man has undertaken any special responsibilities;and where his contract does not involve the manufacture or supply by him of any article or structure.

I recognise that it can be strongly argued that there is no logical basis for drawing a distinction between on the one hand the responsibilities of a contractor who designs and erects a building, and on the other hand the responsibilities of an engineer or the consultant who merely designs it.

Furthermore, there will be cases where the contractor will be liable to the employer for faulty design, and will then wish, in third party proceedings, to recover the sum he has paid to the employer from the independent consultant who in fact designed the building in question.

I have come to the firm conclusion, however, that it is not open to this court, except where there are special facts and special circumstances, to extend the responsibilities of a professional man beyond the duty to exercise all reasonable skill and care in conformity with the usual standards of his profession. There are many authorities which establish that this is the accepted duty of a professional man.'

Defence to fitness for purpose

Although the use of reasonable skill and care will not serve as a defence against failure to achieve fitness for purpose, there are other possible defences.

First, amongst these is the defence that the purpose has not been clearly identified in the contract. Second, is the defence that the works have been used for something other than the intended purpose as stated in the contract.

For static civil engineering projects such as a road or bridge, the accuracy of traffic forecasts and load requirements would be relevant. For a performance contract such as a water treatment works the quality of raw water delivered to the works might be relevant. If in such a case the raw water delivered to the works falls outside the parameters indicated in the design brief, the designer's obligations to achieve a specified quality of treated water may be negated.

This does raise the prospect that in some circumstances the designer may find the liability of fitness for purpose, even where it expressly applies, easier to dispose of than the liability of using reasonable skill and care. This is because the skill and care liability applies to both the information made available in the tender and other non-contractual information the designer should reasonably have known about. And the designer is not excused the obligation to use skill and care because the purpose of the works is uncertain or because there has been misuse.

Lawyers, of course, will be alert to this possibility and in actions for breach arising from design failures where fitness for purpose is claimed to apply will no doubt also plead the alternative case of failure to use skill and care.

4.5 *Design liability in the Design and Construct Conditions*

Although fitness for purpose liability can be implied into design and construct contracts, such liability will not be implied if it is inconsistent with the express terms of the contract. The question in the D and C Conditions is whether there is fitness for purpose liability or whether there are express terms reducing the liability to the lower standard of skill and care.

That question is not answered clearly but there can be little doubt from the explanatory memorandum issued by the Conditions of Contract Standing Joint Committee on release of the D and C Conditions that the intention was to exclude fitness for purpose and substitute skill and care. The memorandum says:

> 'Unless otherwise stated in the Employer's Requirements, the Contractor will *not* be responsible for the suitability or fitness for purpose of the completed Works. To this extent the Contractor is in the same legal position as any outside Consultant retained by the Employer'.

Causes of uncertainty

The uncertainty in the D and C Conditions has a number of causes.

The first lies in the effectiveness of clause 8(2)(a) of the Conditions in excluding liability for fitness for purpose. Clause 8(2)(a) reads as follows:

> 'In carrying out all his design obligations under the Contract including those arising under sub-clause (2)(b) of this Clause (and including the selection of materials and plant to the extent that these are not specified in the Employer's Requirements) the Contractor shall exercise all reasonable skill care and diligence'.

Some lawyers say these words are sufficient to exclude any implied term of fitness for purpose. Others say the words are not a limitation of liability but are simply explanatory of a minimum duty. That is, in carrying out any design obligation, whether it be design of something required to be fit for purpose or not, the contractor is to exercise all reasonable skill, care and diligence.

The second cause of uncertainty on whether fitness for purpose is excluded lies in the combined effect of the wording of clauses 1(1)(e) and 6(2)(a). Clause 1(1)(e) states that the employer's requirements may describe:

> 'the standards performance and/or objectives that are to be achieved by the Works or parts thereof.'

Clause 6(2)(a) states:

> 'The Contractor shall except as may otherwise be provided in the Contract submit to the Employer's Representative such designs and drawings as are necessary to show the general arrangement of the Works and that the Works will comply with the Employer's Requirements'.

If, as seems evident from these clauses, the contractor has an obligation to design and construct works which will comply with required performance objectives, then fitness for purpose far from being excluded from the Conditions, is positively demanded.

The third cause of uncertainty lies in the wording of clause 8(2)(b). The clause says:

> 'Where any part of the Works has been designed by or on behalf of the Employer and that design has been included in the Employer's Requirements the Contractor shall check the design and accept responsibility therefor having first obtained the approval of the Employer's Representative for any modifications thereto which the Contractor considers to be necessary.'

The problem here for the skill and care only line of thought is that the contractor cannot accept responsibility for another's design on a skill and care basis. If the design fails, it is not open to the contractor to raise the defence that he used all reasonable skill and care in undertaking the design.

Clause 8(2)(a) does specifically include design obligations under clause 8(2)(b) in its requirements for the contractor to exercise all reasonable skill, care and diligence, but that requirement presumably applies to checking the employer's design.

Consensus of opinion

It is not in the nature of lawyers to agree where there is scope for disagreement so it cannot be said that there is a firm consensus of opinion amongst lawyers on whether fitness for purpose applies in the D and C Conditions.

Perhaps all that can be said is that of those who have spoken out on the subject there appears to be a large majority with a leaning towards fitness for purpose.

The position in other standard forms

The uncertainty on design liability in the D and C Conditions could have been avoided and is avoided in other standard forms of contract, by different wording.

Thus in JCT 81 the contractor's liability is very clearly limited to that of a professional designer. Clause 2.5.1 reads:

> 'Insofar as the design of the Works is comprised in the Contractor's Proposals and in what the Contractor is to complete under clause 2 and in accordance with the Employer's Requirements and the Conditions (including any further design which the Contractor is to carry out as a result of a Change in the Employer's Requirements), the Contractor shall have in respect of any defect or insufficiency in such design the like liability to the Employer, whether under statute or otherwise, as would an architect or, as the case may be, other appropriate professional designer holding himself out as competent to take on work for such design who, acting independently under a separate contract with the Employer, had supplied such design for or in connection with works to be carried out and completed by a building contractor not being the supplier of the design.'

In the New Engineering Contract the contractor's design liability is fitness for purpose unless option clause M1 applies. That reads:

> 'The contractor is not liable for Defects in the works due to his design so far as he proves that he used reasonable skill and care to ensure that it complied with the Works Information.'

Some forms leave nothing to implied terms and expressly include for both skill and care and fitness for purpose responsibilities. The two do not have to be seen as alternatives and can obviously be complementary in the sense that the contractor has a responsibility to use skill and care in his design in order to achieve fitness for purpose.

Thus the ACA/BPF form says in clause 1.2:

> 'the Contractor shall exercise in the performance of his obligations under this Agreement all the skill, care and diligence to be expected of a properly qualified and competent contractor experienced in carrying out work of a similar scope, nature and size to the Works'.

And in clause 3(1) the contractor warrants that:

> 'those parts of the Works to be designed by the Contractor will be fit for the purposes for which they are required.'

In similar manner, the IChemE Red Book says:

> 'clause 3.2 Subject to the express provisions of the Contract, all work carried out by the Contractor under the Contract shall be carried out with sound workmanship and materials, shall conform to good engineering practice and shall be to the reasonable satisfaction of the Engineer.
>
> clause 3.3 Subject as aforesaid, the Plant as completed by the Contractor shall be in every respect suitable for the purposes for which it is intended.'

If fitness for purpose applies

If fitness for purpose liability does apply in the D and C Conditions what are the consequences? Less perhaps than might be expected from the intensity of the debate.

Unless the matter is raised in connection with insurances or the provision of design warranties it will only be a matter of importance in exceptional circumstances. Principally those where there has been a failure attributable to defects in design but the contractor maintains there has been no negligence in the design. But, as shown in the *IBA* v. *EMI* case and others mentioned earlier avoiding the charge of negligence when there has been failure is no easy matter. And in any event there is the point illustrated in the

Wimpey v. *Poole* case that insurance cover for the contractor may depend on his proving negligence not in his disclaiming it.

4.6 Unforeseen conditions and design liability

The effect of unforeseen conditions on design liability has to be considered in the light of various circumstances:

- whether the standard of liability is fitness for purpose or skill and care
- whether the liability attaches to a contractor with obligations to complete the works or to a professional designer with no obligations beyond design
- whether the unforeseen conditions manifest themselves before or after completion of the works
- whether the meaning of unforeseen conditions is confined to ground conditions or extended by definition, or by interpretation of the courts (see the *Humber Oil* case in Chapter 8), to wider matters
- whether the contract provides for the employer to take the risk of unforeseen conditions.

Professional designers

It cannot be negligent for a professional designer whose liability is limited to skill and care to fail to take account of genuinely unforeseeable conditions in his original design.

Negligence might only arise if the designer fails to modify his design for conditions discovered before completion or fails to warn in respect of conditions discovered after completion.

Fitness for purpose liability

The designer who accepts liability on a fitness for purpose basis, whether he be a contractor designer or a professional designer on special terms, takes on an absolute liability. Relief from the liability is not obtained by proving there has been no negligence.

Therefore, in the absence of other considerations, a designer who undertakes fitness for purpose takes the risk of unforeseen conditions.

The obligation of contractors

Design and construct contractors are burdened with both the obligation to design and the obligation to construct. Consequently, if they encounter unforeseen conditions during construction they have no alternative but to modify their design, as far as is possible, to achieve completion.

Whether or not they receive extra payment depends upon the terms of the contract. If it is a lump sum price, with fitness for purpose and no unforeseen conditions payment clause, the contractor is left with all the costs.

If however, the contractor's design liability is no more than skill and care and there is an unforeseen conditions payment clause, the contractor will be able to recover his costs from the employer.

The position begins to get complicated when there is only skill and care liability but no unforeseen conditions payment clause. For conditions discovered before completion the obligations of the contractor to complete may well take precedence over other considerations and the contractor is probably obliged to modify his design and complete at his own expense. However, for conditions which emerge after completion revealing the design to be inappropriate for those conditions the contractor may not be liable. Providing he has not been negligent in his design much will depend upon whether the defect emerges during the defects correction period or later; and the wording of the defects correction clause, if it still applies.

Unforeseen conditions payment clauses

Most unforeseen conditions payment clauses apply only during the construction of the works so they do not normally come to the assistance of contractors for problems which emerge after completion.

The usual clause which states that the contractor can recover his costs for unforeseen conditions applies whether or not the contractor's liability for design is fitness for purpose or skill and care.

Meaning of physical conditions

Unforeseen conditions clauses are obviously intended to apply to ground conditions and underground artificial obstructions but as discussed in Chapter 8, the phrase 'physical conditions' was given an unexpectedly wide meaning by the Court of Appeal in the *Humber Oil* case. It was held that a transient combination of stress could be a physical condition.

Whether or not that decision would have been reached if clause 12 of the ICE Sixth edition had referred to ground conditions instead of physical conditions may never be known. But with the benefit of hindsight clause 12 of the ICE Sixth edition and its counterpart in the D and C Conditions, neither of which refer anywhere to ground conditions, appear to give more relief to the contractor than was intended.

Unforeseen conditions in the Design and Construct Conditions

Assuming first that the contractor's design liability is limited to skill and care:

- The contractor redesigns as necessary to overcome unforeseen physical conditions encountered during construction and the employer pays the costs under clause 12.
- The contractor will not be liable after completion since his obligation under clause 49(2) is only to rectify at his own cost defects due to his failure to comply with his obligations under the contract. Providing the contractor has used skill and care in his design there is no failure.

Assuming now that the contractor's design liability extends to fitness for purpose:

- For conditions encountered during construction the contractor redesigns and clause 12 applies as above with the employer paying the costs.
- For defects emerging after completion due to unforeseen conditions the contractor would appear to be fully liable.

4.7 Obligations and responsibilities for design in the Design and Construct Conditions

Form of tender

In the form of tender the contractor offers to design, construct and complete the works in conformity with:

- the employer's requirements
- the conditions of contract
- any other listed documents.

Under a contract made by simple written acceptance of tender, the contractor's obligations and responsibilities will, in the absence of any other listed documents, be found in the employer's requirements and the conditions of contract.

Form of agreement

In the form of agreement, the signing of which is not essential to the formation of the contract, the contractor undertakes to design, construct and complete the works in conformity with the provisions of the contract.

The contract by definition includes the contractor's submission so it follows that any obligations and responsibilities in the contractor's submission which are not in conflict with the employer's requirements, which always prevail, will be binding on the contractor.

For further discussion on the implications of this see Chapter 7.

Clause 8(1) – contractor's general obligations

By clause 8(1)(a) the contractor is to design, construct and complete the works:

- subject to the provisions of the contract, and
- save in so far as it is legally or physically impossible.

The point has been made earlier in this chapter that where part of the design has already been undertaken by or on behalf of the employer the contractor's obligation can only be to complete the design notwithstanding that he is required to accept responsibility for the whole of the design.

The subject of impossibility is addressed in Chapters 6 and 19. But in short there is impossibility which can be overcome by change – that is, impossibility falling short of frustration; and there is impossibility which is absolute. The complication in the D and C Conditions regarding the first of these is that the contract, as defined, allows only for changes in the employer's requirements.

This has the effect that any change by the contractor to rectify faults or deficiencies in his submission appears to fall outside the 'contract' and he is then no longer constructing the works in accordance with the 'contract'.

'Reasonably to be inferred from the contract'

The obligation in clause 8(1)(b) that the contractor is to provide all design service, labour, materials etc. of a temporary or permanent nature required for the design, construction and completion of the works 'so far as the necessity for providing the same is specified in or reasonably to be inferred from the contract' has potentially greater significance in the D and C Conditions than the near equivalent provision in the ICE Sixth edition.

In both contracts the contractor has the obligation to provide everything necessary for the works. But whereas under the ICE Sixth edition that does not imply that the contractor stands all the costs since the contract is on a remeasurement basis, under the D and C Conditions when the contract is lump sum, which will be normal, clause 8(1)(b) does imply that the contractor's price includes for everything.

Clause 8(2) – contractor's design responsibility

Clause 8(2)(a) states that the contractor shall exercise all reasonable skill, care and diligence in carrying out all his design obligations.

As shown earlier in this chapter this is the minimum standard the law applies to any designer. The uncertainty in the D and C Conditions is whether this is the only standard which applies or whether the higher standard of fitness for purpose can also apply.

Clause 8(2)(b) – contractor to accept responsibility for the employer's design

Clause 8(2)(b) may prove to be the most contentious clause in the D and C Conditions. It states that where any part of the works has been designed by or on behalf of the employer, and that design has been included in the employer's requirements, the contractor shall:

- check the design
- accept responsibility therefor
- having first obtained the approval of the employer's representative for any modifications considered necessary.

The difficulty with this clause is not so much the principle that the contractor should accept responsibility for the employer's design – some other standard forms such as MF/1 require the same – although there it must be said the contractor does have rights to decline. The real difficulty is that the clause apparently takes effect after the contract has been made and no indication is given as to what happens if the contractor and the employer's representative fail to agree on modifications.

Transfer of responsibility

Considering first the point on timing. When does the contractor accept responsibility for the employer's design?

The explanatory memorandum on the D and C Conditions issued by the Conditions of Contract Standing Joint Committee says this:

> 'Where any part of the design is included in the Employer's Requirements responsibility for that part is transferred to the Contractor when the Contract is awarded, save that the Contractor has the right to check that part of the design and to obtain the approval of the Employer's Representative to any necessary modifications before assuming that responsibility.'

Had clause 8(2)(b) been drafted to accord with such ideas it would perhaps have been more workable. But as it stands, clause 8(2)(b) does not give the contractor a right to check the employer's design; it places a firm obligation on him to do so. And acceptance of responsibility under the clause clearly follows the checking of the design; it does not precede it.

No doubt what the draftsmen of the D and C Conditions had principally in mind was that prior to the award of the contract the contractor would satisfy himself that the employer's design was acceptable. It is unlikely that they intended that only after the contract was awarded would the contractor have to check the employer's design to see if it worked. Perhaps something along these lines should have been said:

'The contractor shall be deemed to have satisfied himself before the award of the contract of the suitability and accuracy of any design undertaken by or on behalf of the employer and included in the employer's requirements and shall upon award of the contract accept responsibility for any such design.'

Approval to modifications

The second part of clause 8(2)(b) requiring the contractor to obtain approval to any modifications considered necessary to the employer's design is also poorly worded.

If indeed modifications are necessary to the employer's design before the contractor can accept responsibility it should not be a matter of the contractor seeking approval but of the employer's representative being obliged to order the necessary changes by way of variation.

In the event of the contractor and the employer's representative failing to agree on whether modifications are necessary two situations can be envisaged. One that the works are constructed without modification and without the contractor accepting responsibility for the employer's design. The other that progress on the works is suspended pending resolution of the dispute. However on this it has to be said that neither conciliation nor arbitration offers the ideal solution but it is not clear what other mechanism could apply.

Cost of checking the employer's design

Clause 8(2)(b) is silent as to which party pays for the costs of checking the employer's design.

From the general obligation in clause 8(1)(b) the cost would appear to fall on the contractor.

Statutory checks

Under clause 8(4) the contractor is obliged to arrange and pay for any statutory checks on the design.

However under clause 26(2) the employer is obliged to make repayment of all fees required to be paid in compliance with statutes and regulations and there is nothing in the wording of clause 26(2) to exclude fees payable under clause 8(4).

Clause 6(1) – further information required for design

Clause 6(1) requires the contractor to give notice in writing if further information is required from the employer for the design or the construction of the works.

Such information will normally be required by way of clarification or amplification of the employer's requirements and in itself it does not give the contractor any entitlement to extra payment. Unless the information can be shown to be a change in the employer's requirements the contractor will be deemed to have allowed for it in his price – clause 8(1)(b).

However, clause 6(1)(b) does entitle the contractor to payment of extra cost and an extension of time for delay if the employer's representative fails to issue any information properly requested within a reasonable period.

As to what is reasonable, that may depend amongst other things upon the detail in the contractor's clause 14 programme and on his rate of progress. Many standard forms of contract make it a specified requirement that programmes should show dates for the supply of further information (in so far as this can be foreseen) but this has not been the policy of ICE contracts prior to the New Engineering Contract and there is no such requirement in the D and C Conditions.

Submission of designs and drawings

Clause 6(2)(a) requires the contractor 'except as may otherwise be provided in the contract' to submit to the employer's representative such designs and drawings:

- as are necessary to show the general arrangement of the works,
- and that the works will comply with the employer's requirements.

Construction to such designs and drawings shall not commence until the employer's representative has consented thereto.

Clearly in a well organised project the employer will have satisfied himself before awarding the contract that he approves of the contractor's general arrangement of the works and that the contractor's designs and drawings comply with his requirements. Those designs and drawings will then be incorporated into the contract and it is not clear what purpose is served by requiring the contractor to seek further consent.

It will, of course, be probable that the contractor will be producing working drawings as the works proceed but note that the clause refers only to designs and drawings 'necessary' to show compliance with the employer's requirements. It is most unlikely that this extends to all of the contractor's designs and drawings but if it does a massive task falls on the employer's representative in examining prior to giving consent and delays would seem to be inevitable.

Withholding of consent

Clause 6(2)(b) empowers the employer's representative to withhold consent to any design or drawing which in his opinion does not comply with the employer's requirements or any other provision of the contract. He is to inform the contractor in writing with reasons and the contractor is required to re-submit the design or drawing with appropriate modifications.

This clause has been much criticised for its potential interference with the contractor's responsibility for design. And, of course, like clause 8(2)(b) it fails to address the very likely matter of disagreement. Moreover the point is made that by clause 1(7)(a) the giving of any consent by the employer's representative does not relieve the contractor of any of his obligations under the contract.

The net effect of this is that the employer's representative can require changes to the contractor's designs and drawings before giving consent but the employer is not liable if the changes prove unworkable and the contractor has to revert to his original designs.

This illustrates a fundamental difference between the D and C Conditions and the ICE Sixth edition. In the ICE Sixth edition the engineer has power and responsibility; in the D and C Conditions the employer's representative has power but very little responsibility.

A point worth noting with regard to clause 6(2)(b) is that it refers expressly to 'such design or drawings'. That is to say it applies only to the designs and drawings required under clause 6(2)(a). However, when it comes to reasons for withholding consent these are extended in clause 6(2)(b) from non-compliance with the employer's requirements to 'any other provision of the contract'.

It might have been expected that any design or drawings which showed compliance with the employer's requirements but not compliance with some other provision of the contract would oblige the employer's representative to make a change in the requirements. But again, as in clause 8(2)(b), the obligation is put on the contractor to make modifications.

Contractor's proposals for change

Clause 6(2)(c) deals in a wholly inadequate way with contractor's proposals for change.

It deals only with design or drawings to which consent has been given and it says only that the employer's representative shall be notified of any proposed changes and that modified designs or drawings shall be submitted for further consent.

It has nothing to say on the valuation of such changes or how changes are to be brought within the matrix of the contract if they deviate from the designs and details in the contractor's submission.

Failure to grant consent within a reasonable time

Clause 6(2)(d) deals with failure by the employer's representative to give consent or notice of withholding consent within a reasonable period following submission or re-submission of designs or drawings. In the event of such failure the contractor is entitled to an extension of time and payment of reasonable costs.

As with clause 6(1)(b), and clause 8(3) there is no express mention of delay; simply failure to respond within a reasonable time. Contractors should note this with interest.

Copyright

Clause 7(1) provides, as would be expected, that the employer retains copyright of his designs and the contractor retains copyright of his. This, of course, is only as between the parties.

Clause 7(1)(b) gives the employer what amounts to a licence to make copies of the contractor's designs etc. for:

- the purposes of the contract, and
- after completion for,
  - operation
  - maintenance
  - dismantling
  - reassembly

- repair
- alteration
- adjustment

of the permanent works.

4.8 *The designer under the Design and Construct Conditions*

The designer of the works

Part 2 of the appendix to the form of tender requires the contractor to state who is to carry out the design work and the elements for which the contractor has not yet appointed a designer.

The Conditions do not preclude in-house design nor do they place any restrictions on the contractor's choice of designer. However, in practice, the employer's requirements may have something to say on these matters.

Clause 4(2)(a) of the Conditions states that the contractor must obtain the employer's consent prior to making any change in the designer named in the appendix.

Note firstly that it is the 'Employer's consent' and not that of the employer's representative. This raises an interesting point on who can actually give the consent.

Regarding the names in the appendix, this presumably applies only to professional firms and not to individual members of an in-house design team.

Professional indemnity insurance

The D and C Conditions do not expressly require professional indemnity insurance for the design of the works and indeed clause 21 relating to insurance of the works is worded so as to exclude the implication that it might be required.

However, many employers will no doubt require such insurance and will include a statement to that effect in the employer's requirements.

There is a point, often made, that only in exceptional circumstances can design insurance be obtained for any liability other than the use of reasonable skill and care and that a designer who takes on fitness for purpose liability may invalidate his policy. In fact the point is sometimes put forward to support the argument that design liability under the D and C Conditions can only be on a skill and care basis – because that is all that can be covered by insurance. But, of course, there is nothing in contract law to say that a contractor only carries those risks he can insure against and patently in the ordinary course of his business he carries many risks which are uninsurable.

Collateral warranties

Collateral warranties, or duty of care letters as some are called, serve two purposes. Firstly, they circumvent the restrictions of the doctrine of privity of contract by extending contractual relationships beyond the close parties to the contract. Secondly, they can extend the range and the duration of liabilities beyond those which apply at common law.

They operate principally to protect employers and others with a commercial interest in a project against failures in the normal chain of contractual liability. Thus, by collateral warranties an employer can obtain direct rights against designers, sub-contractors, suppliers and others engaged in a project.

The terms in such warranties are frequently onerous both as to time and liability. For example, it is not uncommon to require a professional designer to undertake fitness for purpose and to extend liability beyond the normal limitation periods. Not surprisingly such warranties are rejected by many and signed with reluctance by others.

The D and C Conditions say nothing about collateral warranties but again, as with professional indemnity insurance, the employer's requirements may include for them; particularly in connection with professional designers employed by the contractor.

Where warranties are an employer's requirement it is essential that the form of the warranty should be specified.

Chapter 5

Price and payments

5.1 *Introduction*

Whilst working under traditional contracting arrangements the majority of civil engineers have not found it necessary to consider the complexities of the pricing and payment schedules introduced elsewhere in the construction industry over the last decade or so. But the advent of the D and C Conditions and other design and construct forms will certainly change that. The days of remeasurement contracts based on rates may not yet be numbered; but to many civil engineers of the next generation such contracts may be a curiosity.

Most design and construct contracts are executed on a lump sum price basis but, as the popularity of the IChemE Green Book has shown, there is a place and quite strong demand in some sections of the industry for cost reimbursable contracts.

Interim payments

A major effect of the various pricing changes now coming into operation is that the familiar scheme for interim payments based on measurement of work completed up to the end of each monthly period is not appropriate, not necessary, or not even possible for some contracts. Without a priced bill of quantities, the measurement of work items on its own says little about value.

For each design and construct contract, therefore, it is necessary to consider in some detail how interim payments are to be valued and when they are to be made. Some schemes are sophisticated, others exceedingly simple; but generally it can be said that less time and effort is needed in the preparation and checking of interim payment applications whatever scheme is used compared with the monthly grind of the remeasurement system.

Valuation of variations

Another financial change between design and construct contracts and traditional contracts is the way variations are valued. Instead of the basis of valuation being bill rates, variations in design and construct contracts are usually valued with more attention to quotations and cost.

This change in part stems from the probable absence of a priced bill of quantities but it is also related to a wider vision of how contractors should be reimbursed for variations and how certain species of delay and disruption claims arising from variations can be eliminated.

Contract price analysis

It is unusual to find a lump sum price contract which does not require the contractor to submit with his lump sum a contract price analysis.

The analysis may be used by the employer in a number of ways:

- for assessment and comparison of tenders
- as a cash flow indicator
- for determining amounts due as interim payments
- as a guide to the cost of changes in the employer's requirements .

Although the D and C Conditions do not expressly require the submission of a contract price analysis this gap can be filled in the employer's requirements. It would be unwise, if not worse, for any employer to embark on a substantial lump sum contract without the benefit of a contract price analysis.

Arrangements in the Design and Construct Conditions

In summary, the pricing and payment arrangements of the D and C Conditions are as follows:

- Pricing can be lump sum or remeasurement; neither is specified.
- Interim payments can be periodic or activity related according to the particular payment schedule for the contract.
- Variations are to be valued on a fair and reasonable basis by a process of quotation and negotiation.

- There are no express requirements for a contract price analysis or a rates schedule to assist in valuing variations; both are likely to be necessary.
- The Conditions are not suitable for cost reimbursable contracts without substantial amendment.

Unlike the price and payment arrangements in the ICE Sixth edition, the arrangements in the D and C Conditions are not meant to stand alone. A supplementary schedule is necessary stating the price basis of the contract; what information the contractor is to provide; how interim payments are to be made; and how variations are to be valued. The D and C Conditions have the merit of flexibility but it is for the employer and his representative to ensure in the contract documentation that such flexibility does not lead to uncertainty.

5.2 *Firm price or cost reimbursable*

As mentioned above the D and C Conditions are not intended for, and are not suitable for, cost reimbursable contracts without substantial amendment. But that is a point of significance to employers only after they have made the fundamental decision on whether to opt for a cost reimbursable contract in preference to a firm price contract.

Given that a major objective of the transfer of design responsibility to the contractor is to reduce the employer's risk it might seem logical to suppose that the employer would pursue the same policy of risk reduction and always seek a firm price – ideally on a lump sum basis. However, some employers have learned from experience that there are circumstances when a firm price contract is neither the cheapest nor the most risk free option. And, odd though it may seem, a cost reimbursable contract can give greater certainty of price.

At the centre of this conundrum is the point that a firm price contract, whether remeasurement or lump sum, can be overwhelmed by delays or variations and the employer then ends up paying cost-plus but without the benefit of control.

Circumstances favourable to cost reimbursable contracts

- where the project cannot be defined in detail at the start
- where the employer wishes to retain flexibility for change

- where the employer wishes to make a rapid start
- where unusual risks would lead to high contingencies being included in firm price bids
- where the project involves developing technology or state-of-the-art construction which would again lead to high contingencies
- where there are matters of confidentiality which preclude the employer's requirements from being circulated.

Controlling cost reimbursable contracts

It is frequently said in criticism of cost reimbursable contracts that the contractor has no financial incentive to minimise the costs to the employer. If the contractor is on a straight percentage fee that is probably true but not many contracts are now let on that basis.

Most cost reimbursable contracts are now let with fixed fees or target fees and many have guaranteed maximum prices. In fact so stringent are the terms of some cost reimbursable contracts that contractors carry greater financial risk if costs exceed targets than can possibly be allowed for as risk provisions in the fees.

But also essential in a cost reimbursable contract to ensure proper control is a detailed schedule defining precisely what are the reimbursable cost elements and what are the items in the contractor's fee. The IChemE Green Book and the New Engineering Contract both address this issue. The D and C Conditions do not; and they should not be used, even if they are amended, without such a schedule.

5.3 *Lump sum or remeasurement*

The D and C Conditions are drafted to cover both lump sum contracts and remeasurement contracts.

By habit of training, civil engineers may feel more comfortable with remeasurement contracts but the point is made many times in this book that design and construct contracts are inherently lump sum. At its simplest, this is because only the contractor can prepare and warrant the accuracy of the quantities of his own design. And if the contractor warrants the accuracy of the quantities then remeasurement is unnecessary. The principal reason for remeasurement in a traditional contract with engineer's design is that the employer does not warrant the accuracy of the engineer's quantities although he does warrant that they are estimated.

Advantages of lump sum contracts

The principal advantage to the employer of a lump sum contract is that he has greater certainty of price than with a remeasurement contract. The employer takes less risk and the contractor takes more.

Secondary advantages to the employer are:

- bid evaluation is straightforward
- less scope for contractor's claims
- cash flow can be controlled by an interim payment schedule
- the payment scheme may provide an incentive to the contractor to maintain progress and complete activities.

For the contractor, lump sum contracts have advantages despite the burden of additional risk. For example:

- turnover projections can be made with greater certainty than with other types of contract
- cash flow projections can be made similarly reliable
- costs of remeasurement are avoided
- sub-contracts, if let on the same basis, are easier to control
- variations do not expose inadequacies in individual rates
- greater commercial confidentiality can be maintained.

Disadvantages of lump sum contracts

For the employer:

- variations can be expensive
- omission variations may not result in corresponding price savings.

For the contractor:

- increased risks on sufficiency of price
- fewer opportunities for claims for additional payment.

5.4 *Interim payment schemes*

The D and C Conditions endeavour to cover various types of

contract – lump sum and remeasurement – but in doing so inevitably omit the detail required to make anything but the simplest scheme for interim payments work without support from other documents.

Thus clause 60(1) refers to 'the amounts ... due under the Contract' leaving open what this means. The explanation must be sought elsewhere. And clause 60(2) refers to the payment schedule included in the contract – in anticipation that such a schedule exists.

The Conditions clearly contemplate that a payment schedule will be provided in the contract; but as a fall-back provision, clause 60(2)(b) allows for payment on the basis of the amount due 'in the opinion of the Employer's Representative'.

Payment schedules

Various schemes are in operation for interim payments. Some standard forms of contract specify a particular scheme; others like the D and C Conditions leave the matter open and rely on the inclusion in the contract of a payment schedule. Some schemes combine a variety of methods and as an example of this a sample payment schedule for a water treatment plant is included at the end of this chapter.

In brief, it can be said that payments are either made on a periodic basis or a stage payment basis; but even this is something of a simplification because some periodic payments only become due when activities are completed. These are sometimes described as 'milestone' schemes.

Periodic payments

True periodic payments are expressed in contracts as a contractor's right although they are sometimes made subject to the attainment of certain specified minimum values. Such payments may be based on:

- The tender price (or a specified part thereof) divided equally into periodic payments. This is the simplest form of payment scheme but it suffers from the obvious disadvantage that it does not encourage the contractor to maximise progress.

- An assessment of the value of the work done at the end of each period. The ICE Sixth edition uses this scheme with assessment based on measurement. The D and C Conditions contain the outline for a similar scheme as a fall back position where there is no payment schedule but the Conditions do not indicate how the assessment is to be made.
- Progress relative to programme based on a pre-determined expenditure plan. This is the scheme used in the Department of the Environment's standard form GC/Works/1 (edition 3). It relies on a formal approach to monitoring progress and it encourages, if not enforces, the contractor to produce and work to a realistic programme.
- An assessment of work completed against an activity schedule which breaks down the contract price into lesser lump sums. Unlike the milestone scheme described below this scheme does not require an activity to be fully completed before payment for the activity is due.

Stage payments

Stage payment schemes entitle the contractor to payment only when he has reached a specified stage of construction or fulfilled other conditions. For example, completion of foundations on a building project; or delivery of plant to site on a process contract. Such schemes have the advantage of great simplicity but they produce irregular cash flows for both parties. In any event there are limits in large construction projects on how much of the burden of financing the work during construction can be placed on the contractor.

Milestone payments

Milestone schemes are usually something of a cross between periodic payments and stage payments. The contract provides for monthly payments to be made but value can only be included for activities (items of work) which are completed. These schemes need a detailed activity schedule to indicate the payment milestones.

The New Engineering Contract and The Department of Transport design and construct form both use versions of milestone schemes.

5.5 Valuation of variations

Just as there is variety in the interim payment schemes of design and construct contracts there is variety in the rules for the valuation of variations.

The rules fall into four major categories:

- Valuations made with reference to rates in a priced bill of quantities. Used in traditional ICE contracts but not appropriate to lump sum contracts unless the contractor is required to submit priced bills. This does apply in many building contracts. Difficulties with rates based valuations are that they do not always compensate the contractor for his costs and they do not easily accommodate the knock-on effects to other work and the costs of delay and disruption.
- Valuations made with reference to a schedule of rates incorporated into the contract for the purpose of valuing variations. Open to the objection that the rates may be excessive unless tenderers are advised that notional calculations will be made for the assessment of tenders. Useful to have a schedule of rates as a fall back position for quotation based valuations and suitable as such for the D and C Conditions.
- Valuations on a cost-reimbursable basis. This is the method used in the New Engineering Contract in conjunction with quotations. Operates on the basis that the contractor should not be at financial risk from variations but it requires the contract to contain the framework for cost identification and some allowance for the contractor's fee. Payment on a conventional daywork system is an alternative form of cost reimbursement.
- Valuations derived from quotations for each variation. Most design and construct contracts including the D and C Conditions use this method as the starting point for valuations but there have to be supplementary rules to deal with quotations which are unacceptably high. The D and C Conditions fall back to negotiations and then to a fair valuation by the employer's representative.

Logic of the quotation method

Variations in design and construct contracts differ from those in traditional contracts because the contractor carries the responsibility for the effect of the change on the integrity of the design.

Consequently, most variation clauses in design and construct contracts operate on a contractor's consent basis and some forms, such as the IChemE forms and MF/1 expressly entitle the contractor to refuse to perform any variation which he believes will prejudice fulfilment of his obligations under the contract or which exceed in total value a specified percentage of the contract price. The D and C Conditions do not go as far as this, requiring only consultation with the contractor, but in practice the contractor would be unwise to proceed with any variation he was not comfortable with on safety or design grounds.

Cost of preparing quotations

Because the responsibility for variations is placed on the contractor it is entirely appropriate that he should be given the opportunity to submit a quotation for the execution of any variation. The employer can then decide in the light of the quoted cost whether or not he wants to proceed with a change in his requirements. Some contracts expressly entitle the contractor to recover the costs of preparing quotations for variations not proceeded with but the D and C Conditions are silent on this.

Cost of delay and disruption

In civil engineering contracts there is probably nothing more upsetting to an employer than the arrival of a huge claim for delay and disruption with the contractor's final account. And more often than not, the main supporting plank to such a claim is the alleged effect, in terms of delay and disruption, of variations to the works.

To counter this virtually all new standard forms of contract have tightened the rules on the valuation of variations to ensure that the costs of delay and disruption of each variation are considered and included for in the valuation.

The D and C Conditions say only that wherever possible the delay consequences of each variation shall be agreed before the variation is issued or before work starts. The New Engineering Contract, however, takes the matter much further with a programme revision scheme which obliges the contractor to illustrate how any delay or disruption is caused.

Allowing for risk in quotations

Most quotation based valuation methods operate on the principle that once a quotation has been accepted it binds the contractor both in price and any time revision which is included. That is, the contractor takes the risks of sufficiency of price and time allowances.

This is frequently the cause of dispute on the acceptability of quotations. How much in any quotation is a reasonable allowance for risk; and how much is overpricing? This is where a system for building up quotations – such as that in the New Engineering Contract – comes into good effect. Without such a system, all too often the quotation method will fail and the valuation of variations will revert to more time consuming and contentious practices.

5.6 *Sample payment schedule*

General

Payment of the contract price will be made by monthly instalments within the provisions of clause —— of the Conditions of Contract. Each instalment will consist of a payment as appropriate in respect of the following categories subject to the terms detailed in each respective category.

Project management

The amount entered in the Schedule of Prices for project management multiplied by the proportion of the total amount in the Schedule of Prices in respect of sections —— and —— below certified to date by the employer's representative in respect of those sections.

Design

The amount entered in the Schedule of Prices for design in six equal monthly instalments at the end of each of the first six calendar months after the date for commencement.

Supervision of construction

The amount entered in the Schedule of Prices for supervision of construction multiplied by the proportion of the total amount in the Schedule of Prices in respect of sections —— certified to date by the employer's representative in respect of those sections.

Civil engineering works

The amount included in the Schedule of Prices for civil engineering works multiplied by the proportion of work properly completed to date in accordance with the contract, including an allowance for materials on site. Agreement of the proportions complete will be made using the specified progress monitoring system.

Mechanical, electrical, instrumentation and control work

- Items up to £25 000
  The amount entered in the Schedule of Prices for items of mechanical, electrical, instrumentation and control equipment up to the value of £25 000 multiplied by the proportion of the equipment properly installed to date in accordance with the contract.
- Items greater than £25 000
  The amount entered in the Schedule of Prices for items of mechanical, electrical, instrumentation and control equipment of value greater than £25 000 in certified stages as detailed below.

Stage of completion	% Payment of amount in Schedule of Prices
(1) Approval of drawings	10%
(2) 50% of manufacture	65%
(3) Delivery to site	100%

The contractor shall substantiate that the stage has been achieved to the satisfaction of the employer's representative before payment will be certified.

Chapter 6

Allocation of risk

6.1 *Introduction*

Employers faced with making a choice between forms of contract will want to know of any particular contract:

- does it work?
- what are the risks?
- how do the risks compare with other contracts?

For the D and C Conditions it can safely be said that they will work – because civil engineers, being reasonable men, will make them work. Drafting difficulties, here and there, may look formidable but the employer will get his requirements even if a few lawyers earn fees along the way.

As to risks, the employer can be assured that under the D and C Conditions he takes less risk than under traditional forms of contract with engineer's design. He has the benefit of single point responsibility and providing he opts for lump sum pricing he has greater certainty of price than can ever be achieved under remeasurement contracts.

As to the balance of risk and comparison with other contracts, the D and C Conditions are derived from standard conditions long recognised to be amongst the fairest in the industry. There is, of course, a shift in the balance of risk, and that shift is in favour of the employer, but the Conditions generally retain an even-handed approach and are far from oppressive to the contractor.

Some design and construct forms are more favourable to contractors; the IChemE Red and Green Books and the IEE/IMechE MF/1 being obvious examples. But interestingly enough many employers choose these forms simply because, being favourable to the contractor, they are relatively dispute free.

Other design and construct forms, particularly those drafted in recessionary times when the commercial power of the parties favours the employer, strike, on the face of it, a better bargain for the employer with much greater allocation of risk to the contractor – no extensions of time for inclement weather, no payment for unforeseen conditions, no payment for statutory undertakers' delays. Whether or not the employer will continue to get a better bargain in times of economic recovery remains to be seen.

For contractors, whichever design and construct form is used they take more responsibility and more risk than in traditional contracting. But this is not all bad news. Competition should be less cut-throat; the competent will succeed; innovation will be rewarded; and, in short, good contractors will drive out the bad.

6.2 Design risks

The design liabilities of the contractor under the D and C Conditions are discussed in Chapter 4 and the assumption by the contractor of those liabilities is clearly the primary heading for the allocation of risk.

In summary the design risks of the D and C Conditions are:

- The contractor is responsible for the integrity and workability of his design (subject to impossibility).
- The contractor is responsible for design undertaken by, or on behalf of, the employer.
- The contractor warrants that his design will satisfy the employer's requirements. Accordingly, the contractor takes the risk on inadequate design and under–specification.
- The contractor is responsible for the selection of materials and specialist sub-contractors and takes the risk on their suitability.
- The contractor is responsible for the performance of the design team. Therefore, the contractor takes the risk on:
  - late supply of design information
  - mistakes
  - revisions to details
  - co-ordination.

6.3 Pricing risks

The re-allocation of risk between traditional contracting and design and construct contracting comes not only from the transfer of design responsibility itself but from the financial implications of the transfer. Prominent amongst these are risks associated with the contract price.

Quantities

Design and construct contracts are inherently lump sum contracts because remeasurement on bills of quantities is generally out of place. It would be imprudent for the employer to take responsibility for the accuracy of bills of quantities prepared by the contractor; and if the contractor is responsible for the accuracy of his own bills there is no case for remeasurement.

Here lies a major shift in the allocation of financial risk. Under the D and C Conditions, provided they are used on a lump sum basis, the contractor takes the risk on quantities. The provision of traditional ICE Contracts (clause 56(1) of the ICE Sixth edition) that the engineer should determine the value of the works by admeasurement is conspicuously missing from the D and C Conditions.

Method of measurement

Directly associated with quantities is the matter of the method of measurement. In traditional contracts (clause 55(2) of the ICE Sixth edition) errors in the method of measurement are corrected at the expense of the employer. This is a major and a successful source of claims for the contractor. But again in the D and C Conditions the method of measurement is of little relevance. There is no equivalent of the ICE Sixth edition provision for the correction of errors; nor is there any equivalent of the ICE Sixth edition clause 56(2) for adjustment of rates. The contractor therefore takes the risk on his measurement of the quantities.

Contingencies and prime cost items

Employers should not overlook that they take the risk on the adequacy of sums in the tender price for contingencies and prime cost items. This applies even if the contractor enters his own sums – a practice which should not be permitted.

Valuation of variations

The procedures for the valuation of variations under clause 52 of the D and C Conditions do not work to the certain advantage of either party. The intention is that the employer should pay a fair and reasonable price for all variations and this should be based on quotations provided by the contractor.

If anything, however, the change from the traditional method of valuing variations on bill rates is in favour of the contractor because he does not carry the risk of valuations based on bill rates which may be inadequate. Instead the contractor will expect to be paid on what amounts to a cost plus basis. If the contractor has provided a schedule of rates for the valuation of variations this will usually be little more than a daywork schedule specific to the contract and will in effect be cost plus.

Fluctuations

Matters which need to be considered in the allocation of risk on price are which party should take the risk of:

- labour tax fluctuations and other statutory increases
- ordinary price changes after the date of the tender.

In the D and C Conditions, as with most other standard forms, the risk on labour tax matters is placed on the employer. This is the effect of clause 69.

However it will be commonplace to delete this clause so placing the risk on the contractor.

The D and C Conditions do not provide for ordinary price increases except to make the note in clause 71 (special conditions) that if a contract price fluctuation clause is to be incorporated it should be numbered as a special condition.

Unless such a special condition is included the risk of inflationary cost increases lies with the contractor.

6.4 *Construction risks*

Site information

Clause 11 of the D and C Conditions follows the rules of traditional ICE contracts in deeming that the contractor has inspected the site and satisfied himself as to the sufficiency of his price. The contractor is entitled to rely on information provided by the employer, and, in the D and C Conditions as with the ICE Sixth edition, the employer must make available all the information in his possession.

That, however, is not to say that the employer has any obligation to obtain site information. He has no such obligation at common law and none under the Conditions. Thus, as a general rule, the contractor takes the risk on site conditions.

The contract may provide some relief from this, as it does in clause 12 of the D and C Conditions and other ICE contracts. And the employer may choose to undertake site investigations himself rather than leaving the task to individual tenderers. But there is some evidence that in design and construct contracts employers are less disposed to do this, believing, perhaps, that individual tenderers are the best judge of their site investigation requirements and that the costs of site investigation are somehow saved.

In fact, far from saving costs by leaving individual tenderers to make their own investigations, employers end up paying more unless the tenderers band together to share information. Recognising this, a practice is developing whereby the employer funds the joint site investigation of the tenderers.

Adverse physical conditions

There are strong arguments for saying that contractual provisions placing the risk of unforeseen ground conditions on the employer have no place in design and construct contracts.

The basic point is that, ideally, risk should be allocated to the party most able to control the matters involved. This should minimise the consequences.

It is, therefore, logical that clause 12 in the ICE Sixth edition should place the risk of unforeseen physical conditions on the employer, since it is his engineer who is responsible for design. But this argument does not apply to the D and C Conditions with contractor's design. Indeed, if the employer is to pick up the bill for unforeseen conditions, the contractor has no incentive to design a scheme which, by virtue of its robustness, is not inherently susceptible to such conditions.

The second argument used in favour of putting the risk of the unforeseen on the employer is that it is said to be cheaper for the employer to pay for what does happen rather than for what might happen. Whether this is true or not depends very much on economic circumstances and the related ability of contractors to remain competitive with tenders which include an allowance for risk.

The matter can be put to the test, as it is by The Department of Transport's design and construct form, by calling for tenders with and without an unforeseen conditions clause. The employer can than decide in the light of the tender prices whether or not to take the risk or leave it with the contractor.

Although clause 12 of the D and C Conditions is not expressed in any way as an optional provision there have been informal suggestions from the drafting committee that it would not be to the detriment of the Conditions if it were treated as such.

A further argument against the inclusion of clause 12 in the D and C Conditions is that it is likely to operate unfairly, in these Conditions, against the employer. That is, the employer will take all the risk on extra cost but will gain none of the benefits of any cost savings or reduced quantities. This is because the contract is likely to be lump sum with no remeasurement and clause 52 does not deal with the valuation of omission variations. Thus, if rock is found at a higher level than anticipated and that causes the contractor extra costs, the employer pays. But if the higher than anticipated rock level produces cost savings, as well it might, there is no mechanism in the Conditions for passing the savings to the employer. In short, the employer takes the risk of extra costs and the contractor takes the benefit of any savings. This is not how clause 12 works in the ICE Sixth edition and other traditional ICE contracts.

From the employer's point of view, a significant secondary advantage of deleting clause 12 from the D and C Conditions would be that clause 8(5) of the Conditions, making the contractor responsible for the adequacy of his methods of construction, would not be under threat from the decision in the *Humber Oil* case discussed in Chapter 8. In that case, under the ICE Fifth edition, the contractor succeeded in recovering his costs from the employer when his installation equipment collapsed due to unforeseen physical conditions.

6.5 *Impossibility*

At common law, in the absence of express provisions to the contrary, the employer does not warrant that the works can be built. The classic case is *Thorn* v. *Corporation of London* (1876) where the contractor was to take down an old bridge and build a new one.

The design prepared by the employer's engineer involved the use of caissons which turned out to be useless. The contractor completed the works with a different method and sued for his losses on the grounds that the employer warranted that the bridge could be built to the engineer's design. The House of Lords held that no such warranty could be implied.

At common law the contractor who undertakes to construct and complete the works warrants expressly, or impliedly, that he can do so. If he fails he is liable in damages. Thus it was said in *Taylor* v. *Caldwell* (1863):

> 'Where there is a positive contract to do a thing, not in itself unlawful, the contractor must perform it or pay damages for not doing it, although in consequence of unforeseen accidents, the performance of his contract has become unexpectedly burdensome, or even impossible.'

These principles have little effect in traditional ICE contracts where the engineer is usually under a duty to issue variations necessary to complete the works (clause 51 of the ICE Sixth edition) and the contractor is paid on a measurement basis for the work done. But in lump sum design and construct contracts the principles are of greater relevance. In such contracts, the provisions relating to impossibility are of prime importance in so far as risk is concerned.

Impossibility in the Design and Construct Conditions

In the D and C Conditions two clauses refer to impossibility:

- clause 8(1) which states the contractor's obligation to design, construct and complete the works, 'save in so far as it is legally or physically impossible'
- clause 63(1) which states that if circumstances outside the control of both parties arise which render it impossible or illegal for either party to fulfil his contractual obligations the works shall be deemed to be abandoned.

Clause 63(1) does not present too much difficulty. It corresponds generally with the legal principles of frustration and the Conditions provide a payment scheme which broadly lets the loss lie where it falls. The clause deals only with supervening events and its application is likely to be infrequent.

Clause 8(1) on the other hand may prove to be one of the most difficult clauses of the Conditions to interpret and apply. It leaves open the meaning of impossible and it fails to indicate the consequences of impossibility.

As discussed in Chapter 8, the meaning of impossible in clause 8(1) may well be something short of frustration. That is to say, it may be impossible to construct the works in accordance with the contract; but a variation in design would make construction possible. What is not clear is whether the contractor is obliged to vary his design; and even if he does so, how such a variation can be incorporated into the contract. The contract as defined in clause 1(1)(g) of the Conditions allows only the employer's requirements to be varied.

As to the financial consequences when impossibility under clause 8(1) comes into effect, this is not addressed in the Conditions. The contractor would not be in breach of contract; so he would not be liable for damages. But it is questionable whether or not he would be able to retain payments made in respect of part-completed work.

6.6 *Time and delays*

The general principle at common law is that the contractor is obliged to complete on time unless prevented from doing so by the employer. In short the contractor takes the risk of all delays for which the employer is not responsible. The contractor is liable for damages for late completion and his own costs of the overrun.

Standard forms of construction contracts have traditionally reduced the contractor's risk by allowing extensions of time for circumstances beyond the control of the parties – often referred to as neutral events. Typical amongst these are delays due to weather, strikes, unforeseen conditions etc.

The D and C Conditions include various neutral events under clause 44(1):

- certain delays referred to in the Conditions
- exceptional adverse weather
- other special circumstances.

Adverse weather

Exceptional adverse weather has traditionally been included in civil engineering contracts as a ground for extension of time. The contractor is, more often than not, at the mercy of the elements and whilst he is left to take the risk on his own costs he is usually relieved the risk of paying damages for late completion.

However a recent trend in central government contracts, GC/Works/1 (edition 3) and The Department of Transport design and construct form, is to omit adverse weather from extension of time clauses. That leaves the risks of weather wholly with the contractor.

This is not necessarily unfair because there is another trend developing – to provide for completion on time by paying the contractor to accelerate. In effect, paying the contractor to forgo his right to an extension of time. Considered in this way it is not illogical to omit adverse weather from extension provisions. Adverse weather is not normally a reimbursable event as an extension of time as is any delay caused by the employer. And whilst it is right that the employer should pay for completion on time if he causes delay, it is a different matter to say that the employer should pay the contractor to complete on time if the weather causes delay.

6.7 Statutes and approvals

In traditional contracting it is usually left with the contractor to serve notices and with the employer to obtain statutory consents. This is in keeping with the employer's responsibility for design.

In design and construct contracting the position is potentially different. Obtaining statutory consents may be dependent upon the detail of the contractor's design. It is not unreasonable for some, if not all, of the risk of obtaining consents to be allocated to the contractor.

The D and C Conditions take a median approach to this. Clause 26(3) requires the contractor to ascertain and conform to all regulations and bye-laws but the employer is left with responsibility for obtaining planning permission 'unless the contract otherwise provides'.

Planning permission

In design and construct contracts the risk of obtaining planning permission would seem to rest naturally with the contractor who, by his design, either complies or fails to comply, with the relevant restrictions. But that, of course, is a simplification.

Firstly, there is the matter of outline planning permission. Hardly a matter to be left unresolved at the time of tender. Only the bravest contractor, given the responsibility of obtaining outline planning permission, would go into contract not knowing the outcome of his application. It does happen occasionally in the building industry but it is difficult to see it happening in civil engineering.

The responsibility for obtaining detailed planning consent is, however, a different matter. It should be possible to indicate to the contractor at the time of tender what planning restrictions apply. The contractor can then be left to use his design skills to satisfy both the employer's requirements and the planning authorities.

The danger for the employer in retaining responsibility for obtaining all planning permissions, as he does in the D and C Conditions, is that the detailed scheme is not his but the contractor's – unless, of course, the employer's requirements are rigidly defined. Consequently the employer may be taking a risk which he cannot control – if for example his requirements are expressed only as performance parameters – and the contractor has freedom in design. The risks for the employer in such circumstances are considerable since, amongst other things, the employer warrants in clause 26(3)(c) that all planning permissions will be obtained in due time.

6.8 Other contractors and statutory undertakers

In the D and C Conditions, as in traditional ICE contracts, the contractor is not entitled to exclusive possession of the site; and clause 31 requires the contractor to afford reasonable facilities to other contractors and statutory bodies. The contractor can recover from the employer costs not foreseeable at the time of tender in complying with this.

By clause 31 and other provisions of the Conditions the employer takes on much of the risk of the performance of other contractors and statutory undertakers both in terms of extra cost and delay. This may not be unreasonable but it is not essential. It is possible in design and construct contracts, to draft the employer's requirements and the contractor's scope of work so that the contractor controls and takes responsibility for other contractors and statutory undertakers. The risk then lies with the party in control.

Chapter 7

Definitions, documents, consents and notices

7.1 *Introduction*

This chapter examines:

clause 1 – definitions and interpretation
clause 5 – contract documents
clause 68 – serving of notices.

Clause 1

Clause 1 follows the pattern of other ICE contracts in providing meanings for key words and phrases used in the text. Twenty-one items are so defined – a few less than in the ICE Sixth edition because 'specification', 'drawings', 'bill of quantities' and 'tender total' are either not applicable or are part of the employer's requirements or the contractor's submission.

All the defined items, except the word 'cost' commence with capital letters in the text as a means of identification. By a minor printing error, the word 'works' in paragraph 3 of the form of agreement appears without such capitalisation and for consistency this should be corrected when the form of agreement is used.

Clause 1 contains the usual aids to interpretation – that words importing the singular include the plural; that headings and marginal notes are not to be taken into consideration in the construction of the contract; and that references to clause numbers are to clauses in the Conditions.

Clause 1(6) expressly provides that communications which the contract requires to be in writing may be handwritten, typed or printed and sent by hand, post, telex, cable or facsimile.

The major difference between clause 1 of the D and C Conditions and clause 1 of the ICE Sixth edition is the inclusion in clause 1(7) of the D and C Conditions of provisions relating to consents and approvals given by the employer's representative. The ICE Sixth edition has similar provisions in respect of the engineer but they are attached to other clauses and are not presented with the same collective force.

Clause 5

In the D and C Conditions, clause 5 covers both the resolution of ambiguities or discrepancies and the supply of contract documents.

These are broadly the same matters as covered in clauses 5 and 6 of the ICE Sixth edition although there is considerable re-adjustment of the provisions with different emphasis on the obligations of the parties.

Clause 68

There is no change between the ICE Sixth edition and the D and C Conditions in clause 68 on the serving of notices.

7.2 *Identification definitions*

Employer

The employer is to be named in the appendix to the form of tender. Care should always be taken to ensure that the correct legal title of the employer is inserted and that the title corresponds with that to be used in the letter of acceptance and the form of agreement. Many a contractual dispute is clouded by issues of identity and the complications caused by having one name for the employer in the appendix and another name in the award of the contract can take a good deal of unravelling.

Clause 1(1)(a) states that the employer 'includes the Employer's personal representatives successors and permitted assigns'.

The latter pair usually present little difficulty since to take effect as the employer various legal or administrative formalities have to be undertaken including, in the case of assignment, obtaining the consent of the contractor under clause 3 of the Conditions.

The personal representatives of the employer are, however, in a different category. They may be the legal replacements of the employer using the phrase in a restricted legal sense. But the intention of the clause may indicate that other employer's personal representatives have the capacity to act for the employer. Such representatives could be liable for the employer's contractual obligations. Nevertheless, this is a difficult area as shown by the case of *Sika Contracts Ltd* v. *Gill* (1978) in which a chartered civil engineer was held personally liable for payment on a contract where he was acting both as agent of the employer and in his professional capacity. Mr Gill's mistake was to have accepted a quotation from Sika on his own headed notepaper.

The D and C Conditions do go a long way towards eliminating problems with personal representatives by having a defined and named employer's representative to act for the purposes of the contract. This is a considerable improvement on the building design and build form, JCT 81, which leaves 'the employer' with contractual duties to fulfil including granting extensions of time and ascertaining loss and expense.

Contractor

The contractor means the person, firm or company to whom the contract has been awarded. As with the employer, the definition in clause 1(1)(b) includes personal representatives, successors and permitted assigns.

With contractors, take-overs and liquidations often bring problems of identity not least because the identity of a limited company lies not in its name but in its registration number.

The phrase 'to whom the Contract has been awarded' in clause 1(1)(b) leaves open the method of award. The form of tender contemplates award by acceptance of the contractor's submission but as discussed later in this chapter there is a strong case in design and construct contracts for a more formal approach based on signing the form of agreement.

Employer's representative

The role of the employer's representative is examined in Chapter 9. The main comment to make here is that clause 1(1)(c) restricts the appointment to a 'person'.

In other ICE contracts the engineer is a person, firm or company although admittedly in the ICE Sixth edition a person must be named to act on behalf of the engineer. But this still permits the employer to appoint a firm of consulting engineers and it is that firm with its resources and its professional indemnity insurance on which the employer is entitled to rely.

In the D and C Conditions the employer appears to be prevented from appointing a firm although the practical case for doing so will often be greater than in traditional contracting. One way around this difficulty is for the employer to enter into a contract with a firm to provide an appropriate person; a simpler way may be to amend the Conditions.

The person appointed by the employer as his representative may either be named in the appendix to the form of tender or notified to the contractor in writing within seven days of the award of the contract and before the commencement date in accordance with clause 2(2).

Clause 1(1)(c) permits the employer to replace his representative from time to time and unlike some other design and construct forms of contract such as MF/1 there is no requirement for the employer to seek the contractor's consent before doing so.

Contractor's representative

The contractor's representative may, or may not be, the contractor's site agent. Provision exists for both in the Conditions but clause 15(3) permits the representative to be the site agent.

In so far as the contractor's representative is not the site agent his position is new to ICE conditions of contract. However it is essential that where the contractor's obligations extend beyond construction to design the contractor should have an overall person in charge. This is recognised in most standard forms of design and construct contracts and has wisely been followed in the D and C Conditions.

Clause 1(1)(d) defines the contractor's representative as the person appointed by the contractor to act for the purposes of the contract. His appointment is to be notified in writing to the employer and he can be replaced from time to time by the contractor.

Perhaps by an oversight in drafting neither clause 1(1)(d) nor clause 15 makes the appointment itself a contractual requirement. The implications of this are discussed in Chapter 9 under the role of the contractor's representative.

7.3 *Contract document definitions*

The contract

Clause 1(1)(g) defines the contract as including:

- conditions of contract
- employer's requirements
- contractor's submission
- written acceptance of the contractor's submission
- such other documents as expressly agreed
- contract agreement (if completed).

Traditional ICE contracts refer to:

- conditions of contract
- specification
- drawings
- bill of quantities
- tender
- written acceptance of the tender
- contract agreement (if completed).

The employer's requirements and the contractor's submission clearly take in the specification, drawings and bill of quantities (if any). The contractor's submission becomes the tender.

Nowhere in the Conditions is any order of precedence given to the contract documents except in clause 5 where the employer's requirements are to prevail over the contractor's submission.

Conditions of contract

It is accepted practice in civil engineering contracts that the conditions of contract are incorporated by reference into the contract and this will no doubt continue with usage of the D and C Conditions.

One effect of the practice is that amendments to standard conditions are made in separate documents. In the building industry and elsewhere this rarely happens because the standard conditions often preclude extraneous amendments. So any changes are made in the pages of the standard conditions and are initialled by both parties. This has a number of benefits:

- it imposes discipline on amending
- amendments are conspicuous
- the amendments are literally within the standard conditions.

Civil engineers might well consider the merits of this, but otherwise, if standard conditions are amended care is needed to ensure that the amendments become effectively incorporated. It is probably best in the D and C Conditions to keep amendments separate from the special additional conditions which should be numbered from clause 71 onwards and usually deal with price fluctuations, corruption, restrictions on publicity and the like.

Employer's requirements

By definition in clause 1(1)(e) the employer's requirements are those requirements identified as such at the date of award of the contract and any subsequent variations.

This definition not only expressly permits changes to the requirements after the award of the contract but it also implies that changes may be made between submission of the form of tender and the award. This flexibility is essential to preserve the status of the employer's requirements as the basis of the contractor's obligations.

The scope of the employer's requirements is left open and they are not necessarily confined to technical matters. Although clause 1(1)(e) uses the phrase 'which may describe the standards performance and/or objectives' the phrase is preceded by the word 'and' which indicates that there may well be other matters.

Contractor's submission

By definition in clause 1(1)(f) the contractor's submission is his tender offer together with such modifications or additions which may be agreed prior to the award of the contract.

The Conditions offer no obvious scope for variation of the contractor's submission after the award – a point discussed in some detail in Chapters 3 and 16. In so far that the submission is a financial sum this rigidity is understandable but in connection with technical aspects of the submission it may not be desirable or practical.

7.4 Formation of the contract

Single stage or two stage

Design and construct contracts do not fit easily with the routine familiar to engineers of forming contracts by a simple letter of acceptance of a tender. The form of tender of traditional ICE contracts encourages this by its paragraph:

> 'Unless and until a formal Agreement is prepared and executed this Tender together with your written acceptance thereof, shall constitute a binding Contract between us.'

But traditional contracts are by their nature suitably formed by the single stage process of tender plus acceptance. With design and construct contracts, however, a stage of clarification, negotiation and modification between submission of tender and award of the contract is commonplace. The nature of many design and construct contracts is that they are not properly single stage nor are they fully two stage in the sense that stage 1 is for design only and stage 2 is for construction.

This can produce some difficulty in the public sector where contract procurement rules may preclude the negotiation process or may require equality of approach to all tenderers prior to the award.

In the D and C Conditions the form of tender suggests single stage in its paragraph on acceptance which is identical to that in the form of tender for the ICE Sixth edition. However the definition of contractor's submission with its reference to modifications and additions agreed between the parties prior to the award clearly suggests at least a modest movement towards two stage.

At one time the drafting committee of the D and C Conditions considered including a fairly elaborate code for acceptance within the Conditions but eventually came down in favour of flexibility. Although this approach has its advantages, users of the Conditions need to be aware of the potential pitfalls, including particularly the use of letters of intent.

Letters of intent

Letters from employers to tenderers may be intended to convey:

- acknowledgment

- acceptance of tender
- award of contract
- notice of likely award of contract
- instruction to commence preliminary work
- opening of negotiations
- counter-offers.

Not surprisingly with such variety of intentions, letters of intent need to be expressed with utmost clarity. The Building Law Reports testify to the high numbers of failures.

Disputes which reach the courts often centre on the contractor's right to payment for work done on 'contracts' which never proceed as intended but a greater danger perhaps with the D and C Conditions is that a contract could be formed prematurely with important matters still unresolved. Thus under the D and C Conditions a written response to a contractor's submission which indicates acceptance would, in the light of the wording of the form of tender, form a binding contract. The effect would be to fix the contractor's submission at that point and to make any subsequent changes to the employer's requirements into variations. Any subsequent negotiations, instead of being part of an acceptance procedure would be post-contract matters with entirely different financial consequences.

Form of agreement

Another pitfall for users of the D and C Conditions is that the relationship between the form of agreement and the form of tender is not the same as that in the ICE Sixth edition and other versions of ICE contracts.

At first sight the differences are not apparent. The D and C Conditions and the ICE Sixth edition contain an identical clause 9 which requires the contractor to enter into and execute a contract agreement 'if called upon so to do'. This suggests that the contract agreement is something of an optional extra in both sets of Conditions. And, of course, with the ICE Sixth edition the form of agreement is seen principally as the device to extend the legal limitation period from the statutory six years to the twelve years which applies to contracts executed as a deed (previously called contracts under seal).

This remains one of the uses of the form of agreement under the D and C Conditions but the agreement has another important use in defining the obligations of the parties.

With the ICE Sixth edition the form of agreement simply reflects the obligations arising from the tender and its acceptance. With the D and C Conditions both the employer's requirements and the contractor's submission may change between tender and award of contract. Consequently the obligations covered by the form of agreement may well be different from those envisaged in the form of tender.

A further more complex point is that in the form of tender in the D and C Conditions the contractor offers to complete the works in conformity with the employer's requirements, the conditions of contract, the details in parts 1 and 2 of the appendices and such other documents as listed. In the form of agreement the contractor covenants to complete the works in conformity in all respects with the provisions of the contract which, by definition, includes the contractor's submission. If the contractor's submission offers more than the employer's requirements actually demand, the contractor is apparently covenanting to provide more. Thus in the form of agreement the contractor's obligations may be greater than those in the form of tender.

Clearly both parties need to give more thought to the effect of the form of agreement in the D and C Conditions than they may previously have done in traditional ICE contracts.

7.5 Financial definitions

The D and C Conditions do not have the range of financial definitions found in the ICE Sixth edition. 'Bill of Quantities' as one might expect is omitted but so also is 'Tender Total' which in the ICE Sixth edition is either the total of the bill of quantities at the date of award of contract or the agreed estimated total value at that date.

No name is given in the D and C Conditions to the sum in the form of tender and to overcome that difficulty clause 10 on performance security describes the amount of the bond by reference to 'the estimated total value of the Works at the date of the award of the Contract'.

In practice, users of the D and C Conditions will no doubt find it more convenient to refer to the amount in the tender as the tender sum or the tender total and whether or not that amount be the figure first submitted or the figure agreed after negotiations there should be little difficulty in understanding what is meant.

Contract price

Clause 1(1)(h) describes the contract price as 'the sum to be ascertained and paid for the design construction and completion of the Works in accordance with the Contract'.

This appears to mean the total amount due to the contractor providing he fulfils his obligations under the contract. But the point is not as clear as it might seem. None of the clauses of the Conditions refers to the 'Contract Price' and the only place the phrase is to be found in the contract is in the form of agreement as the consideration payable by the employer.

There is an interesting change in the definition of 'Contract Price' between the ICE Sixth edition and the D and C Conditions. In the ICE Sixth edition the contract price is 'the sum to be ascertained and paid in accordance with the provisions'. This is generally taken to exclude sums payable to the contractor for extra-contractual claims. But in the D and C Conditions the phrase 'in accordance with the provisions' is missing. Perhaps it is intended, therefore, that the contract price should include such sums. Whether or not this extends to deductions for liquidated damages is a matter for conjecture.

Prime cost item

Clause 1(1)(i) describes a 'Prime Cost Item' as an item which contains a sum referred to as Prime Cost which will be used for the supply of goods etc.

The important words here may be 'which will be used' for the rest of the definition is circular and can only acquire meaning within the context of clause 58(2).

Contingency

In the D and C Conditions the phrase 'Contingency' replaces the more familiar 'Provisional Sum' of traditional ICE contracts.

That word change apart, the definition of 'Contingency' in clause 1(1)(j) is very close to the provisional sum definition in the ICE Sixth edition. If there is a change of any significance it is that a 'Contingency' may be used 'in accordance with the specific requirements stated' whereas a 'Provisional Sum' may be used in the ICE Sixth edition 'at the direction and discretion of the Engineer'. The point is discussed further in Chapter 18.

In both sets of Conditions the phrase 'which may be used' appears – indicating the discretionary nature of the use of such sums but it is, of course, a matter of some importance as to who exercises the discretion.

Cost

Clause 1(5) gives the meaning of cost as used in the Conditions. It covers:

- all expenditure properly incurred or to be incurred on or off the site
- design costs
- overheads
- financing charges including loss of interest
- other charges properly allocatable.

Cost does not include any allowance for profit.

For further comment on cost see Chapter 17 of this book and section 2.12 of the author's book on the ICE Sixth edition entitled 'A User's Guide'.

7.6 Works definitions

The D and C Conditions define permanent works, temporary works and the works with the same circular phraseology of the ICE Sixth edition. 'Permanent Works' mean permanent works; 'Temporary Works' mean temporary works; and the 'Works' means 'the Permanent Works and the Temporary Works and the design of both'.

The intention presumably is that the words should be given their ordinary meaning so that permanent works are those which remain on completion and temporary works are those which are removed before completion. This distinction however fails to cater for items which are in the nature of temporary works but are left in place on completion – sheet piling, grouting and cofferdams are possible examples. Fortunately the application of the phrases in the context of the clauses in which they occur is usually fairly obvious. Thus clause 7(1)(b) refers to 'Permanent Works' in relation to post-completion maintenance etc; and clause 14(6) links 'Temporary Works' with methods of construction and the use of contractor's equipment.

Clearly what cannot be done in the D and C Conditions is to seek a distinction between permanent works and temporary works on the basis of who is responsible for design since the contractor is given responsibility for both. This is a significant difference from the ICE Sixth edition where the engineer is generally responsible for the design of the permanent works and the contractor is responsible for the design of the temporary works.

The Works

It is worth noting that it is only the definition of the 'Works' in clause 1(1)(m) which includes design. The intention of this is that when the phrases 'Permanent Works' or 'Temporary Works' are used separately, or together, the design function is excluded. Thus:

- Clause 15(3) requires the contractor's agent to superintend the construction of the 'Permanent and Temporary Works'.
- Clause 21(1) requires insurance of the 'Permanent and Temporary Works'.
- Clause 29(3) relates to the unavoidable consequences of carrying out the 'Permanent and Temporary Works'.

But the converse of this must be that where the word 'Works' is used it does include the design function. Can then the contractor encounter physical conditions in his design office (clause 12)? Must the contractor include design operations in his programme (clause 14)? These are just two of the interesting questions which emerge; there will inevitably be others.

7.7 Time related definitions

Commencement date

Clause 1(1)(n) states simply that the commencement date is as defined in clause 41(1). That clause provides three alternative methods of fixing the date.

Note that it is not the 'Works Commencement Date' as in the ICE

Sixth edition; although by clause 41(2) the contractor is required to start the 'Works' as soon as reasonably practicable after the commencement date.

Certificate of substantial completion

Clause 1(1)(o) defines a certificate of substantial completion as a certificate issued under clause 48. That clause states the procedure and the conditions to be met for the issue of such a certificate.

The D and C Conditions have followed the ICE Sixth edition in using the phrase 'substantial completion'. This is clearly the correct phrase since clause 48 expressly contemplates the possibility of outstanding works. Engineers using the ICE Fifth edition were often dismayed to find they had to issue a 'completion' certificate with works patently still outstanding.

Defects correction period

This is another phrase which follows the ICE Sixth edition. The traditional phrase 'maintenance period' is now seen to be something of a misnomer and is being systematically dropped from civil engineering contracts.

The length of the defects correction period is to be stated in the appendix to the form of tender – usually 12 months.

Clause 1(1)(p) defines the date from which the period begins to run; namely the date on which the contractor becomes entitled to a certificate of substantial completion for the whole of the works or any section or part.

A contract may therefore have numerous defects correction periods in force either simultaneously or concurrently. There will, however, be only one defects correction certificate.

Defects correction certificate

The explanation for the single defects correction certificate is found in clause 61(1) which relates the certificate to the last of the defects correction periods where there is more than one such period.

Clause 1(1)(q) does no more than define the defects correction certificate as the certificate issued under clause 61(1).

Sections

Although a section of the works will usually relate in practical terms to a physical part, the contractual significance of defining a section relates to time.

This is because the contractor is only obliged to finish before the time for completion of the whole of the works any section which is stated in the appendix to the form of tender to have its own time.

Clause 1(1)(r) confirms, therefore, that a 'Section' means a part of the works separately identified in the appendix.

Parts

There is no definition in the D and C Conditions for a part; nor is there in the ICE Sixth edition.

A part of the works which is finished early may, if the contractual provisions are fulfilled, attract its own certificate of substantial completion and defects correction period. But the employer will not normally have a right to occupy or to use a part which is not defined as a section before completion of the whole; Nor will the employer have a right to liquidated damages for late completion of a part not so defined.

7.8 The site

Clause 1(1)(s) defines the site as:

- the lands and other places on, under, in, or through which the works are to be constructed
- any other lands or places provided by the employer for the purposes of the contract
- such other places as may be designated in the contract as part of the site
- such other places subsequently agreed by the employer's representative as forming part of the site.

The definition is essentially the same as in the ICE Sixth edition and the same problems of interpretation and practical application remain. Do the words 'provided by the Employer' apply only to the words 'any other lands or places' which they immediately follow or also to the words 'the lands and other places' which open the clause? The question is relevant to the employer's obligation under clause 26(2) to pay the rates on the site and the contractor's entitlement to payment under clause 60(1)(c) for goods and materials not yet delivered to the site.

The implication from clause 42 which requires the employer to give possession of the site is that the employer has control of the site and that suggests the phrase 'provided by the Employer' in clause 1(1)(s) should be given the wider application.

That view is supported by the absurd result of taking literal application of the words 'on under in or through' in the clause without reference to provision by the employer. It would mean, for example, that land above a tunnel was by definition part of the site.

The provision in clause 1(1)(s) for the site to be extended by agreement of the employer's representative should be used sparingly. It may be necessary to do so for variations which involve work outside the site boundary but otherwise there will rarely be any benefit to the employer.

7.9 Contractor's equipment

The definition of contractor's equipment in clause 1(1)(t) is identical to that in the ICE Sixth edition.

Both sets of Conditions are now in line with practice elsewhere in construction and process engineering in describing the contractor's equipment as things used to construct the works and plant as things intended for incorporation in the works.

7.10 Consents and approvals

Clause 1(7) represents an interesting innovation in ICE contracts in bringing into the section on definitions and interpretation a collection of provisions on consents and approvals.

Some other standard engineering forms, FIDIC Fourth edition and the IChemE books for example do have something to say on consents and approvals in their opening clauses but it is only that consents and approvals shall not be unreasonably withheld. This, however, is not part of clause 1(7) although it might usefully be added as an amendment.

The purpose of clause 1(7) seems to be to pick up provisions which appear in the ICE Sixth edition in specific clauses but have no corresponding place in the D and C Conditions. And also perhaps to emphasise that the agency role of the employer's representative must not be misunderstood as conferring consent or approval by the employer which might then amount to relaxation or variation of the terms of the contract.

Consent or approval not to relieve the contractor of his obligations

Clause 1(7)(a) states that consent or approval by the employer's representative shall not relieve the contractor of any of his obligations or his duty to ensure the correctness, accuracy or suitability of the matter which is the subject of the consent or approval.

This appears to derive in part from clause 7(7) of the ICE Sixth edition which relates to the engineer's approval of permanent works designed by the contractor.

The implications, however, in the D and C Conditions of this provision go well beyond anything which could be contemplated in the ICE Sixth edition. The main concern is that under clause 8(2)(b) the contractor is to take responsibility for any design in the employer's requirements and the contractor requires the approval of the employer's representative to any modifications he thinks necessary. It would appear that the employer's representative can give his approval and the employer can then wash his hands of his responsibility for his own design and any modifications approved on his behalf. The point is discussed in more detail in Chapter 4.

One point to note on clause 1(7)(a) is that it is additional to and not in replacement of the general provision in clause 2(1)(b) that the employer's representative has no authority to relieve the contractor of any of his obligations under the contract.

Failure to disapprove does not prejudice power to take action subsequently

Users of the ICE Fifth and Sixth editions will readily recognise clause 1(7)(b) of the D and C Conditions. They will know it as clause 39(3) of those contracts relating to the removal of unsatisfactory work and materials.

Clause 39 of the D and C Conditions does not extend as far as sub-clause (3). Instead the provision that failure by the employer's representative or any of his assistants to disapprove does not prejudice his power subsequently to take action appears as clause 1(7)(b).

Contractors often see the provision as unfair and obviously late disapproval can lead to cost and delay which might have been avoided. But the employer's representative (or engineer) supervises on behalf of the employer not the contractor and his contractual duty to the employer is to secure proper performance of the contract.

Acceptance of a programme does not imply that the programme is feasible

Clause 1(7)(c) has been developed from clause 14(9) of the ICE Sixth edition but it places greater emphasis on the employer's rights as to time.

Clause 14(9) of the ICE Sixth edition says only that approval of a programme or consent to proposed methods shall not relieve the contractor of any of his duties or responsibilities. Clause 1(7)(c) of the D and C Conditions says that acceptance by the employer's representative of any programme shall not imply that the programme is feasible or appropriate and that the employer's right as to time under the contract shall in no way be impaired by such acceptance. One detects behind this a long history of programme related claims in traditional civil engineering contracts, many arising from over optimistic shortened programmes. Perhaps it is in response to the view expressed in some quarters that the case of *Glenlion Construction Ltd* v. *The Guinness Trust* (1987) discussed in Chapter 8, is of no relevance to ICE contracts.

But it may simply be that in the D and C Conditions the employer's representative has to respond in writing to the contractor's programme and clause 1(7)(c) is confirming the point that if the employer's representative makes a misjudgment or mistake – say by accepting a programme showing longer than the time allowed – the contractor gets no benefit from that error.

7.11 Clause 68 – notices

Clause 68 is identical to its counterpart in the Sixth edition. It deals with notices to be given to the contractor and notices to the employer.

Clause 68(1) provides that all notices to be given to the contractor shall be served in writing at the contractor's principal place of business, and for a company at its registered office.

Clause 68(2) similarly provides that notices to the employer shall be served at the employer's last known address, or in the event of the employer being a company, at its registered office.

Clause 68 applies only to notices to be given to the parties and not to representatives of the parties. So fortunately most of the notice requirements of the Conditions are not brought within the scope of the clause. It applies principally to formal notices such as clause 64 and 65 on defaults and clause 66 on disputes.

Chapter 8

Obligations and responsibilities of the parties

8.1 Introduction

The express obligations and responsibilities of the parties under the D and C Conditions are listed in appendices at the end of this chapter.

Commentary on these obligations and responsibilities is given in various chapters and only the 'General obligations' section of the Conditions (clauses 8 to 19 inclusive) is dealt with in detail in this chapter.

Similarities with the ICE Sixth edition

The following 'general obligations' clauses of the D and C Conditions have much the same wording and effect as corresponding clauses in the ICE Sixth edition:

clause 9 – contract agreement
clause 10 – security for performance
clause 11 – provision of information, inspection of site and sufficiency of tender
clause 12 – adverse physical conditions
clause 16 – removal of contractor's employees
clause 19 – safety and security.

Omissions from the ICE Sixth edition

The D and C Conditions do not use the following 'general obligations' clauses of the ICE Sixth edition:

clause 13 – instructions of the engineer
clause 17 – setting out.

Comment is made later in this chapter and elsewhere in this book on the significance of these omissions.

Modifications to the ICE Sixth edition

The following 'general obligations' clauses are changed from the ICE Sixth edition to take account of the contractor's design responsibility:

clause 8 – contractor's general obligations
clause 14 – programme and methods of construction
clause 15 – contractor's superintendence and representative
clause 18 – boreholes and exploratory excavation

General comment

In setting out the express obligations of the parties, the draftsmen of the D and C Conditions have succeeded in minimising the changes from the ICE Sixth edition – and in some cases perhaps the change is too minimal. Nevertheless this approach has the advantage that users of the Conditions who are familiar with the ICE Sixth edition will feel immediately comfortable in finding that many clauses are unchanged and that many important clauses, such as clauses 11, 12 and 14, have been retained with only marginal change.

8.2 *Clause 8(1) – the contractor's general obligations*

Clause 8(1) expands on the contractor's general obligations to design, construct and complete the works as stated in the form of tender and the form of agreement.

The clause says that the contractor shall:

- subject to the provisions of the contract, and,
- save in so far as it is legally or physically impossible
- design, construct and complete the works, and
- provide all design services, labour, materials, contractor's equipment etc required for design, construction and completion
- so far as the necessity for providing the same is specified in or can be inferred from the contract.

There are two changes of substance from the corresponding clause in the ICE Sixth edition:

- the inclusion of design responsibility
- the limitation 'save in so far as it is legally or physically impossible'.

'Subject to the provisions of the contract'

The phrase 'subject to the provisions of the Contract' in clause 8(1) can be taken to have two meanings. One, that the contractor shall comply with the requirements of the contract; the other that the contractor's obligations may be modified by provisions in the contract. Both are probably applicable.

Compliance is an obvious matter but the modification of obligations is less straightforward. Some cases present no difficulty; for example where the contractor has express rights of termination, as under clause 40(2) for prolonged suspension, or under clause 64(4) for employer's default. Then he is no longer under an obligation to complete the works. But in other cases the modification of obligations may be related to the receipt of instructions. For example if the works are damaged by an excepted risk, what obligation, if any, does the contractor have to rectify and complete unless he receives instructions under clause 20(3)?

'Legally or physically impossible'

The transfer of the phrase 'legally or physically impossible' from its place in clause 13 of the ICE Sixth edition to clause 8(1) in the D and C Conditions opens up some interesting contractual issues. The context in which the phrase is used is quite different between the two contracts.

In the ICE Sixth edition, clause 13 operates in conjunction with clause 51 to require the engineer to issue variations and instructions to overcome impossibility so far as that is possible. The contractor is obliged to comply with those variations and instructions so as to complete the works.

In the D and C Conditions, with no clause 13 and a differently worded clause 51, the 'impossibility' phrase may operate simply as a limitation on the contractor's obligations. It is far from certain that there is any obligation on either party to originate any variation necessary to enable the works to be completed.

Meaning of 'legally or physically impossible'

The meaning of the phrase 'legally or physically impossible' as it is applied in clause 13 of the ICE Fourth edition was considered in the case of *Turriff Ltd* v. *Welsh National Water Development Authority* (1980).

In that case problems arose with tolerances for precast concrete sewer segments and eventually the contractor abandoned the works on the grounds of impossibility. The employer argued that impossibility meant absolute impossibility without any qualifications and since there was no absolute impossibility the contractor was in breach. To sustain this argument the employer contended that the contractor would be obliged to depart from the specification if that was necessary to overcome the impossibility.

The judge declined to accept that impossibility meant absolute impossibility and he held that the works were impossible in an ordinary commercial sense. On the proposition that the contractor was obliged to depart from the specification the judge made the point that if absolute impossibility was the test for impossibility then an equally strict construction of the contract would apply. So what was physically impossible would relate to the precise limits of the specification and that could not be relieved by a deviation on the part of the contractor.

The following extracts from the judgment of His Honour Judge William Stabb QC show how the arguments of the case were examined. The judge said:

> 'But the real issue is as to the construction of the words "physically impossible" in the light of those authorities which establish the Common Law position. [Counsel for the employer] urges me to construe it as having an absolute meaning, so as to give it certainty. "Physically" is coupled with "legally" and he contends that as "legally impossible" must be absolute in the sense that there can be no degrees of legal impossibility, so "physical impossibility" should be read in the same way. Furthermore, he contends that if the contractor finds it absolutely impossible to comply with his obligation to execute, complete and maintain the works in strict accordance with the Contract as stated in Clause 13 of the Contract ("Contract" meaning the General Conditions, Specification, Drawings, Bill of Quantities, Tender and Contract Agreement) then the contractor, in pursuance of his overall obligation to execute and complete the works, is excused from strict adherence to the provisions of the Contract

i.e. he is not only entitled but is required to break the Contract in order to render possible what was otherwise impossible, and in that event he will be excused such a breach or, by virtue of Clause 13, such a departure from the Specification will not constitute a breach.

As to the word "impossibility", I was referred to various dictionary definitions, which in the end show, as one would expect, that something which is physically impossible is something which cannot be done according to the laws of nature. In one sense, this could mean absolutely impossible in that, from the material or physical point of view, the end result could not be achieved whatever the steps taken to try and get round or over the difficulty which was said to render the performance physically impossible. On the other hand, it could be construed as "impossible to perform in compliance with the Specification and Drawing" which is part of the Contract. If that be the correct interpretation, [Counsel for the employer] contends that a contractor, when faced with a physical impossibility resulting from the employer's design or Specification, is under an obligation to render possible that which the employer had by Contract made impossible, and, it would seem, that the contractor must re-design or re-specify some other mode of working, not in compliance with the Contract, in order to achieve the end result.'

'[Counsel for the employer] contends that the absolute meaning avoids uncertainty, but it seems to me that if such a construction be right, the contractor, when faced with what, from all practical points of view, appears to be an absolute impossibility, would still be faced with the uncertainty of knowing whether there was not some means of getting round the obstacle. It must be very rare to find some instance in which human ingenuity, coupled with wholesale deviation from the specified contract works, cannot achieve completion.

One can readily understand the Sale of Goods Act type of case, where the subject matter of the Contract turns out to have become non-existent. In such instances, no Contract has come into existence. But here, there is an express provision for excusal in the event of physical impossibility, and what [Counsel for the employer] contends is that given its absolute meaning, manufacturing and laying were not beyond the bounds of possibility and jointing was rendered possible by relaxation of the Specification. Alternatively, if manufacturing and jointing was impossible to achieve in accordance with the Contract, then Clause 13 would excuse a breach of, or non-compliance with the Contract on the part of the Contractor if he had to resort to such means in order to overcome the impossibility. I am bound to say that I find such a rigid construction difficult to accept, although the linking of legal with physical impossibility makes it difficult to avoid. I think that the answer must be that if absolute impossibility is to be the test, then equally a strict construction of the Contract must also be applied, in the sense that what is physically impossible in accordance with the precise limits of the Specification and Drawings, which are expressly made part of the Contract, should not be permitted to be considered possible, because it can be rendered such by a deviation from the Contract on the part of the contractor.'

Overcoming impossibility within the D and C Conditions

In the ICE Sixth edition scope clearly exists within the contract, by clauses 13 and 51, for overcoming impossibility if alternatives exist. In the D and C Conditions there is a difficult obstacle.

The 'contract' as defined in clause 1(1)(g) of the D and C Conditions includes the employer's requirements and the contractor's submission. The employer's requirements can be varied but not the contractor's submission which is fixed at that 'prior to the award of the Contract'. In short, any change made by the contractor to his submission is outside the contract unless verified by a change in the employer's requirements. Without this latter change the works would no longer be constructed in accordance with the contract. And this, of course, would be a breach.

For an example of the difficulty, consider a contract in which the employer's requirements say only that a bridge should be built from point A to point B to carry a specified traffic loading. The contractor makes his submission and the contract is awarded. The contractor then realises that his design is fatally flawed; it is impossible to build the bridge in accordance with his submission. The Conditions would seem to suggest in clause 6 that the contractor is obliged to put forward a revised design; but the contractor might argue that under clause 8 he is under no such obligation and that his obligations to construct the bridge are at an end. There would be no case for a change in the employer's requirements to resolve the problem.

It may take some years of usage of the Conditions or some drafting amendments to bring anything like certainty into the meaning of clause 8(1).

Provide all services and resources

Clause 8(1)(b) states the contractor's obligation to provide all that is necessary for the design, construction and completion of the works.

The corresponding clause in the ICE Sixth edition rarely attracts much comment because the engineer decides what is in the works and the contractor can argue through the measurement and valuation clauses about his entitlement to be paid.

In the D and C Conditions, however, with lump sum pricing, and the difficulties described above as to what constitutes the contract, clause 8(1)(b) may well become a focus of attention. There will almost certainly be debate on whether anything which is in the contractor's submission in excess of the employer's requirements becomes a contractor's obligation on the grounds that it can be reasonably inferred from the contract.

8.3 Clause 8(2) – contractor's responsibility for design

Comment on the contractor's design responsibility is made in Chapter 4. In this section the detail of clause 8(2) is given with only the briefest comment.

Clause 8(2)(a)

Clause 8(2)(a) requires the contractor to exercise all reasonable skill, care and diligence in:

- carrying out his design obligations under the contract
- carrying out design obligations arising under clause 8(2)(b)
- the selection of materials and plant to the extent they are not specified in the employer's requirements.

This clause is probably intended to exclude fitness for purpose responsibility in design but a great many lawyers have expressed doubts on whether it achieves this.

Clause 8(2)(b)

Clause 8(2)(b) states that where any part of the works:

- has been designed by, or on behalf of the employer, and
- that design has been included in the employer's requirements, then
- the contractor shall check the design, and
  - accept responsibility therefor,
  - having first obtained approval of the employer's representative for any modifications the contractor considers to be necessary.

The potential difficulties of this clause are all too obvious. It is likely to be regarded as the most onerous and difficult clause in the Conditions.

8.4 *Clause 8(3) – quality assurance*

Design and construct contracting is well suited to the implementation of quality assurance. Its characteristic of single point responsibility avoids the difficulties of applying the principles of BS 5750 to organisations with partial responsibility and avoids the problems of interface between organisations.

The D and C Conditions allow for quality assurance but do not make it mandatory except 'To the extent required by the Contract'. The Department of Transport design and construct form in contrast firmly requires the contractor to operate a quality assurance plan for the performance of the contract with plans for design, examination and construction. Moreover, in that contract, the designer of the works must have BS 5750 recognition and similarly sub-contractors should meet the requirements of BS 5750 or operate a quality system of similar standard.

Purpose of quality assurance

Quality assurance is about giving confidence. Applied to construction that should mean less involvement by the employer in supervising the contractor. It remains to be seen in practice how this works out. There is evidence that on some design and construct projects site supervision has been reduced to a token presence but on others the resident engineer and his team of assistants are still employed.

The danger of calling for quality assurance and then seeking to supervise and control is the creation of confusion. And here, perhaps, the D and C Conditions can be criticised. Clause 8(3) requires the contractor to obtain the prior approval of the employer's representative to his quality plan and procedures before each stage of design and construction is commenced.

Precisely how this is intended to operate is unclear but if the contractor has BS 5750 certification why should further approval be needed and what is the position if that approval is withheld?

Provisions of clause 8(3)

Clause 8(3) contains four separate provisions on quality assurance:

- to the extent required by the contract, the contractor shall institute a quality assurance system
- the contractor's quality plan and procedures shall be submitted to the employer's representative for prior approval before each stage of design or construction
- if the employer's representative fails to notify the contractor within a reasonable period of approval or withholding of approval the contractor is entitled to be paid his reasonable costs and to be granted an extension of time
- compliance with an approved system shall not relieve the contractor from any of his duties, obligations or liabilities under the contract

'To the extent required by the Contract'

Clause 8(3) of the Conditions opens with the sentence 'To the extent required by the Contract the Contractor shall institute a quality assurance system'.

The 'Contract' as defined in clause 1(1)(g) includes both the employer's requirements and the contractor's submission so both parties have the opportunity to specify what is required. The employer's requirements may simply call in general terms for quality assurance plans; the contractor's submission may provide the detail.

But whether or not 'the extent required by the Contract' is specified by one party or both there is excellent guidance to be found in CIRIA Special Publication, SP84, 1992 – Quality Management in Construction – Contractual Aspects.

8.5 *Clause 8(5) and clause 19 – safety*

Clause 8(5) and clause 19 of the D and C Conditions both relate to safety obligations. Clause 8(5) covers the contractor's obligations in respect of design and methods of construction; clause 19 deals with the responsibilities of both parties to keep the site safe and secure.

Clause 8(5) – contractor responsible for safety

Clause 8(5) of the Conditions states that the contractor shall take responsibility for:

- the safety of the design
- the adequacy, stability and safety of all site operations
- the adequacy, stability and safety of all methods of construction.

This is similar to clause 8(3) of the ICE Sixth edition except that responsibility for design is added and the contractor is to take 'responsibility' not 'full responsibility'.

The omission of the word 'full' from the D and C Conditions recognises perhaps, that neither under the contract nor by statute is responsibility for safety allocated solely to the contractor. Under the contract the employer has express safety responsibilities under clause 19 and has associated responsibilities under clauses 20 and 22 in respect of 'excepted risks' and 'exceptions'. Moreover as shown below, clause 12 on unforeseen conditions can in some circumstances transfer responsibility from the contractor to the employer.

Under statute, responsibility falls on whoever the Act deems it should fall and private contractual arrangements purporting to state otherwise are of no effect. Thus under the Health and Safety at Work Act 1974 the duties under the Act apply to 'any person' and notwithstanding its wording clause 8(5) of the Conditions does not in any way diminish the responsibility of the employer under the Act.

This may be a point of particular importance to design. For although under clause 8(2)(b) the contractor accepts responsibility for the employer's design, this can only be – as are all contractual provisions – effective as between the parties themselves. It is probable, therefore, that if the employer's design is unsafe, the employer will remain legally responsible.

The effects of clause 12

In the case of *Humber Oil Trustees Ltd* v. *Harbour and General Works (Stevin) Ltd* (1991) under the ICE Fifth edition, the Court of Appeal had to consider whether the contractor's obligation under clause 8(2) affected his rights of claim under clause 12 for unforeseen conditions. The contract was for the construction of mooring dolphins and the contractor selected and used a jack-up barge equipped with a fixed crane. As the crane was lifting and skewing a concrete soffit member it became unstable and collapsed. The contractor submitted a claim under clause 12 that he had encountered physical conditions which could not have been foreseen. The employer disputed this claim, of which more is said elsewhere in this chapter, and argued that even if the contractor did encounter such conditions he had no claim under clause 12 by virtue of the provisions of clause 8(2). The clause stated that 'The Contractor shall take full responsibility for the adequacy stability and safety of all site operations and methods of construction'.

The Court of Appeal rejected the employer's argument that as clause 8(2) was unqualified there was no room for the operation of clause 12. Lord Justice Nourse said:

> 'I cannot construe clause 8(2) as applying to a case where the inadequacy or instability is brought about by the contractor's having encountered physical conditions within clause 12(1). That I think was the instinctive view of the arbitrator, who dealt with this question very briefly. Like the judge, I have not found it an easy question, but like him [...] I would on balance decide it in favour of the contractors.'

The decision in the *Humber Oil* case will cause great concern to employers. It shows that clause 12 cuts across the general principle that the contractor should be free to select his own methods of working but in doing so he must take responsibility for them. It means that the cost to the employer of the contract price will be dependent on how susceptible the contractor's chosen methods of working are to unforeseen conditions. This will not be acceptable to many employers.

The answer is to amend clause 8 or to delete clause 12.

Clause 19 – safety and security

Clause 19(1) on safety and security requires the contractor to have full regard for the safety of all persons entitled to be on the site and to maintain the site in a safe and secure condition. Clause 19(2) places a similar obligation on the employer in respect of any work he carries out on the site with his workforce or any other contractors he employs on the site.

8.6 Clause 9 – contract agreement

Clause 9 requires the contractor 'if called upon' to enter into and execute a contract agreement in the form annexed to the Conditions. The employer is responsible for the cost of preparing the contract agreement.

Without a formal written agreement the legal limitation period for the contract cannot be extended from the standard statutory six years to the 12 years which applies to contracts executed as a deed (alternatively called contracts under seal).

An interesting legal position would apply if the contractor was called on to enter into a contract agreement and refused to do so. Refusal would clearly be a breach of contract – but would the employer have any effective remedy?

Point of formation of contract

There is little doubt in the minds of most civil engineers that a tender and a letter of acceptance constitute a binding contract. The standard form of tender used for the D and C Conditions and for recent editions of other ICE contracts appears to be quite definite on the point: 'Unless and until a formal Agreement is prepared and executed this Tender together with your written acceptance thereof, shall constitute a binding Contract between us'.

However it is quite common, particularly with lump sum contracts, for the letter of acceptance to be no more than a letter of intent – subject to completion of all appropriate formalities. In such cases the exact point of formation of the contract, described in the D and C Conditions as 'the award of the Contract' is not always as clear as it needs to be.

8.7 Clause 10 – performance security

Clause 10 deals with 'performance bonds'.

It states that if the contract requires the contractor to provide security for performance, it shall be provided by a body approved by the employer and shall be in the form of bond annexed to the Conditions.

The appendix to the form of tender shows whether or not a bond is required and the amount, if so required, expressed as a percentage of the estimated contract price. Clause 10(1) limits this to 10%.

Non-provision of bond

If a bond is required it is to be provided within 28 days of the award of the contract. Unlike many standard forms of building contract, there is no express provision in the D and C Conditions for the employer to determine the contractor's employment if the bond is not provided. However, the breach may be sufficiently serious to justify common law determination and support for this is to be found in a South African case, *Swartz & Son (Pty) Ltd* v. *Wolmaransstadt Town Council* (1960).

Payment for cost

Clause 10 requires the contractor to pay the cost of the bond unless the contract provides otherwise.

The model form of bond

Bonds differ considerably in their drafting and the conditions under which they can be called in for payment. At one end of the scale there are 'demand' bonds which can be called in without proof of default or proof of loss; at the other end there are performance or 'conditional' bonds which can only be called upon with certification of default by the contract supervisor and proof of loss.

The model form of bond in the D and C Conditions is a performance or conditional bond and although it does not expressly require a certificate of default, it does rely on both proof of default and proof of loss. It is very similar to the model form of bond used in traditional ICE contracts.

Release of bond

The model form of bond provides a surety which remains in force until the defects correction certificate. Most contractors will offer, as an alternative, a bond which expires on the issue of the certificate of substantial completion. The reason is that the longer the bonds remain in force, the more of the contractor's bonding facility is used.

Disputes on bonds

Clause 10(2) supports the provision in the standard form of bond for arbitration on any dispute regarding the date of the defects correction certificate. The resolution of such a dispute takes place outside the scope of clause 66 and is said by clause 10(2)(b) to be without prejudice to any dispute under that clause.

8.8 Clause 11 – provision of information and sufficiency of tender

Clause 11 covers two closely related matters – the provision of information on the site by the employer and the obligation of the contractor to inspect and examine the site to ensure the sufficiency of his rates and prices.

The corresponding clause in the ICE Sixth edition has attracted much attention and comment and early revisions are expected. But this is not so much because the clause places an express obligation on the employer to provide information – a provision previously lacking in ICE contracts, but because the 'deemed' provisions of the clause are thought by lawyers to be confusing.

Clause 11 of the D and C Conditions, as it stands, is almost the same as the ICE Sixth edition clause – deemed provisions included – but it also places an obligation on the employer to make available all known information on pipes and cables in, on, or over the ground.

Information – general legal position

The general position on responsibility is that, in the absence of specific information and express provisions in the contract to the contrary, the contractor takes all risks from adverse ground conditions. In the old case of *Bottoms* v. *York Corporation* (1892), where neither the contractor nor the employer took boreholes on a sewerage scheme adjacent to a river, it was held that the contractor was not entitled to abandon the contract for difficulties in dealing with water in the excavations since the employer had given no guarantee or representation on the nature of the ground.

In most construction contracts, however, the employer does supply some information and there are usually some express provisions relating to that information and the balance of risk for unforeseen ground conditions. The legal position in such circumstances on the accuracy or adequacy of information given can be exceedingly complex. However, some pointers to the way the law works can be gained from the following cases:

- in *Pearson & Son* v. *Dublin Corporation* (1907) the size of an undersea wall which the contractor was to use was incorrectly marked on drawings by the engineer who had taken no steps to verify the accuracy of his dimensions. It was held that a clause in the contract requiring the contractor to satisfy himself on all dimensions could give protection only against honest mistakes but the engineer's conduct amounted to fraudulent misrepresentation.
- in *Morrison-Knudsen* v. *Commonwealth of Australia* (1972) a dispute arose on a civil engineering contract with the requirement that the contractor should inform himself as to site conditions and a disclaimer on site information provided by the employer. The contractor brought an action against the employer for breach of a duty of care by supplying information that was false, inaccurate and misleading. On the facts the court declined to say that the employer had a duty or care or that the contractor had any cause of action but on the matter of the contractor's obligation to inform himself it was said:

  'The basic information in the site information document appears to have been the result of much highly technical effort on the part of a department of the [employer]. It was information which the [contractors] had neither the time nor the opportunity to obtain for themselves. It might even be doubted whether they could be expected to obtain it by their own efforts as a potential or actual bidder. But it was indispensable information if a judgment were to be formed as to the extent of the work to be done in making the landing strips of the proposed airport.'

- in *Bacal Construction (Midlands) Ltd* v. *Northampton Development Corporation* (1975) on a dispute in a design-build contract as to the accuracy of ground condition information supplied by the employer, it was held that there was an implied term that the ground conditions would be as indicated and that the employer was liable for the cost of the re-designed work.

Clause 11(1) – provision and interpretation of information

By clause 11(1), the employer is 'deemed' to have made available to the contractor, before submission of the tender, all information 'obtained by or on behalf of the Employer from investigations undertaken relevant to the Works' on:

- the nature of the ground and subsoil
- hydrological conditions
- pipes and cables in, on or over the ground.

Lawyers have warned that the word 'deemed' gives perverse interpretation of clause 11(1) – namely that although the employer may provide nothing at all, whatever he has got is deemed to have been given. But this is obviously not the intention of the clause.

Clause 11 does not place any obligation on the employer to carry out site investigations. That is a matter for the employer's commercial judgment. And, as discussed in Chapter 6, some employers will see it to be in their interest to take the initiative in obtaining site information and passing it on to tenderers, whilst others will leave site investigations to the tenderers.

All the clause requires is that the employer should make available whatever information he has. Failure to do so will be a breach of contract for which the employer may be liable in damages.

Interpretation of information

The second provision of clause 11(1) is that the contractor shall be responsible for the interpretation of all information 'for the purposes of the Works'.

Such purposes are no doubt intended to cover design and construction and not the end use of the works.

Clause 11(2) – inspection of site

By clause 11(2) the contractor is deemed:

- to have inspected and examined the site and its surroundings and available information
- to have satisfied himself 'so far as is practicable and reasonable' on:
  - the form and nature of the ground and subsoil
  - the extent and nature of the work
  - the materials necessary for construction
  - the means of communication to the site
  - the access to the site
  - the accommodation required
- to have obtained all information on risks, contingencies and other circumstances which may affect his tender.

'So far as practicable and reasonable'

The proviso 'so far as practicable and reasonable' recognises the difficulties tenderers face in the time available to them and the restrictions they face in carrying out their investigations. However, it is only in respect of ground conditions that the contractor gets express relief by way of payment in appropriate circumstances through clause 12. For other matters, including those listed, the contractor normally takes full responsibility.

Clause 11(3) – basis and sufficiency of tender

Clause 11(3)(a) says that the contractor is deemed to have based his tender on the information made available by the employer and on his own inspection and examination. In the event of conflict between these two it is not clear which is intended to prevail but the contractor has no obvious right of claim if he uses the employer's information in preference to his own and underprices his tender. Indeed, clause 11(3)(b) on sufficiency of price suggests that the contractor should allow for the greater obligation if uncertainty exists.

Correctness and sufficiency of rates

The provision in clause 11(3)(b) that the rates and/or prices shall, (unless otherwise provided in the contract) cover all the contractor's obligations under the contract may prove to have greater effect in the D and C Conditions than in traditional ICE contracts. This is because the great escape route for contractors from the burden of sufficiency of price – clause 55(2) of the ICE Fifth and ICE Sixth editions allowing for correction of errors in the method of measurement – is omitted from the D and C Conditions. And, of course, such a clause has no relevance in lump sum pricing.

The contractor may still, under the D and C Conditions, look to the bracketed phrase in clause 11(3)(b) – 'unless otherwise provided in the Contract' as offering scope for extra payment – which indeed it does as shown in Chapter 17 – but its application will be confined to specific circumstances and will not extend to measurement.

8.9 Clause 12 – adverse physical conditions

Clause 12 provides that the employer rather than the contractor should take the risk of delay and extra cost arising from physical conditions which could not have been foreseen by an experienced contractor.

The suitability of such a provision in a design and construct contract is discussed in Chapter 6. In this chapter, interpretation of the wording of clause 12 and its application is considered.

Clause 12 in the D and C Conditions is much the same as in the ICE Sixth edition except that it is for the contractor and not the employer's representative to say how adverse conditions should be dealt with.

Effect of clause 12

In *Holland Dredging (UK) Ltd* v. *Dredging & Construction Co. Ltd* (1987) Lord Justice Purchas in the Court of Appeal likened clause 12 to a shield protecting the contractor against the provisions of clause 11. By that he meant that, without clause 12, the provisions in clause 11 for rates and prices to cover all obligations would prevail in all circumstances, foreseeable or not. Indeed it was just as well for the contractors that the Court of Appeal took the view that it did, for the judge at first instance in the case had held that clause 12 applied only to supervening events – that is, those arising after the contract was formed,not those in existence at the time of tender.

Meaning of 'physical condition'

Clause 12 applies when the contractor encounters 'physical conditions (other than weather conditions or conditions due to weather conditions) or artificial obstructions which could not have been foreseen by an experienced contractor'.

Most engineers asked to define the meaning of 'physical conditions' in this context would point to something tangible – boulders in clay; a high water table; running sand; hard rock. Most would take it for granted that clause 12 applies to conditions which are found in the ground. Both views need to be re-assessed following the Court of Appeal decision in *Humber Oil Trustees Ltd* v. *Harbour and General Works (Stevin) Ltd* (1991).

In that case, the question arose whether the contractor had encountered physical conditions within the scope of clause 12. As a 300 tonne crane on a jack-up barge was placing precast soffit units on piles the barge became unstable and collapsed causing extensive damage to the works, plant and equipment. The barge was a total loss and had to be replaced. There was much delay and extra cost.

The contract was under the ICE Fifth edition Conditions and the contractor claimed under clause 12 that the collapse of the barge and its consequences was due to encountering physical conditions which could not have been foreseen by an experienced contractor. The dispute went to arbitration.

The arbitrator gave an award in favour of the contractor finding that, although the soil conditions were foreseeable, clause 12 contains no limitation on the meaning of 'physical condition'; that a combination of strength and stress, although transient, can fall within the term; and that in this case an unforeseeable condition had occurred.

The employer appealed maintaining that the question should not be whether the collapse could have been foreseen, which it clearly could not, but whether physical conditions could reasonably have been foreseen. The judge upheld the arbitrator's award but gave leave to appeal. The arguments advanced for the employer before the Court of Appeal would certainly have found favour with many engineers – namely that a physical condition is something material, such as rock or running sand, and that an applied stress is not a physical condition nor is it something which can be encountered. The Court of Appeal, however, rejected the submissions and dismissed the appeal.

Clause 12(1) – notice of conditions

Clause 12(1) provides that the contractor shall give written notice to the employer's representative when he encounters physical conditions or artificial obstructions which in his opinion could not have been foreseen.

This is a requirement which applies whether or not the contractor intends to go on to make a claim and its purpose seems to be to put the employer's representative on notice of potential difficulties.

'Other than weather conditions'

The exclusion for conditions due to weather conditions is taken to apply to immediate weather conditions rather than to the climatic changes which influenced geological strata. But it is still difficult to find a precise boundary.

Artificial obstructions

Artificial obstructions are usually taken to be buried, man-made objects and clause 12 can, in appropriate circumstances, serve as an effective claim clause for statutory undertaker's apparatus which is found to be out of position. However, much depends on the accuracy of location indicated by any information supplied. A general statement that a gas main exists in a road suggests that finding the gas main cannot be unforeseen; whereas finding a gas main in the east footpath after being informed it was in the west footpath could be unforeseen. In *C J Pearce & Co.* v. *Hereford Borough Council* (1968) the precise location of an old sewer, which the contractor had to cross, was unknown. The sewer collapsed into the contractor's excavation. It was held that, even had the contractor claimed under clause 12, which he did not, his claim would have failed since the condition could have been foreseen.

'Could not reasonably have been foreseen'

There are differences of opinion on whether the foreseeability test is wholly objective or whether it allows for the special knowledge of a particular contractor.

In order to put all tenderers on the same footing it is suggested that the test should be wholly objective and that the phrase 'in his opinion' entitles the tenderer/contractor to take the test as expressed literally.

Clause 12(2) – intention to claim

Clause 12(2) requires that if the contractor intends to claim additional payment or extension of time he shall inform the employer's representative in writing when giving notice of encountering unforeseen conditions 'or as soon thereafter as may be reasonable'.

The condition or obstruction encountered must be specified.

Clause 12(3) – measures being taken

Clause 12(3) requires that when giving notice under clause 12(1) or 12(2) or as soon as practicable thereafter, the contractor should give details of:

- the anticipated effects
- the measure being taken or proposed
- estimated costs
- anticipated delay
- anticipated interference.

Such details are to include when appropriate:

- consideration of alternative measures, and/or
- methods of procedure with comparative estimates of costs and delays.

The requirement for these details is a departure from the ICE Sixth edition to recognise the contractor's design responsibility.

Clause 12(4) – action by the employer's representative

The employer's representative, unlike the engineer in the ICE Sixth edition, is not expressly empowered to give instructions on how adverse physical conditions should be dealt with. His express powers are:

- to give written consent to the contractor's measures with or without modification
- to order a suspension under clause 40 or issue a variation order under clause 51.

It might appear that issuing a variation amounts to the same thing as giving an instruction but in the D and C Conditions a variation is no more than a change in the employer's requirements and responsibility for design of that change rests with the contractor. Thus it is not for the employer's representative to instruct how the change should be implemented.

The ordering of a suspension under clause 40 does not necessarily entitle the contractor to recover his costs – the suspension may be necessary for the proper execution or for the safety of the works.

Clause 12(5) – conditions reasonably foreseeable

Under clause 12(5) the employer's representative is required to inform the contractor in writing as soon as he reaches any decision that the conditions could have been foreseen by an experienced contractor. No time limit is set on reaching a decision.

Clause 12(6) – delay and extra cost

The contractor's entitlement to payment under clause 12(6) of the D and C Conditions has reverted back to the policy of the ICE Fifth edition which amounts to:

- the reasonable cost of additional work
- the reasonable cost of additional contractor's equipment
- a reasonable percentage addition for profit to additional work and equipment
- an extension of time for any delay

and

- the reasonable cost of any delay and disruption to the rest of the works (but excluding profit).

8.10 *Clause 14 – contractor's programme and methods of construction*

The D and C Conditions retain the requirement of traditional ICE contracts for the contractor to submit a programme and details of his methods of construction within a specified time of the award of the contract.

Some of the main functions of the programme are of lesser importance in design and construct contracting than in traditional contracting because the employer's representative does not have the same duties as the engineer in respect of the supply of information and co-ordination. But the programme remains necessary in connection with possession of the site, monitoring progress and assessment of delays. Moreover under some payment schemes for lump sum contracts the programme regulates the amounts of interim payments.

The D and C Conditions have not followed the lead of other recent standard forms in requiring the programme to be fully resourced and backed by detail of safety plans and the like. Nor is there any express requirement that the contractor should work strictly to his programme and update his programme at regular intervals to ensure that this can be maintained. However, although the D and C Conditions do not go as far as the New Engineering Contract in establishing a firm link between the valuation of variations and the programme, there is, in clause 52 of the Conditions, reference to valuation of delay which, to operate effectively, requires that the contractor work to a realistic programme.

Shortened programmes

Although the contractor's obligation under the D and C Conditions regarding completion is the same as under other ICE contracts – that is, to complete 'within' the time allowed, there should be less use of the practice of submitting shortened programmes.

Contractors have for many years believed that by submitting programmes showing completion in a shorter time than that allowed in the contract they improve their claims prospects.

It is certainly true that in traditional contracting the shorter the time for completion, the more pressure there is on the employer and the engineer in undertaking their obligations and duties, and the greater the likelihood that acts of prevention will delay the contractor. But this situation does not apply to anything like the same extent in design and construct contracting and in any event the decision in the building case of *Glenlion Construction Ltd* v. *The Guinness Trust* (1987) has done much to dent the confidence of contractors in the effectiveness of shortened programmes as vehicles for claims.

It was held in the *Glenlion* case that there was no implied obligation on the employer to supply information, through his architect, so as to enable the contractor to finish early. And it was said that it was neither reasonable nor equitable that the contractor should be able to place, after the contract had been made, a unilateral obligation on the employer.

Status of programmes

Because the clause 14 programme only comes into existence after the contract has been made it is not a contract document. And so, although it is a breach of contract not to submit a clause 14 programme it is not, in itself, a breach to fail to follow the programme. Failure to proceed with due expedition and without delay is a breach under clause 41(1); but that is a different matter.

It is, of course, possible to draft specific references in conditions of contract to the contractor's programme and that way to create contractual obligations. The D and C Conditions do this in clause 42(2) in relation to possession of the site. But as in the ICE Sixth edition this is the only clause which refers to the clause 14 programme and then, it should be noted, it places obligations on the employer rather than on the contractor.

The fact that the employer can be bound by the contractor's programme is something frequently overlooked by those who show an enthusiasm for tender programmes to be made contract documents. The dangers were well illustrated in the case of *Yorkshire Water Authority* v. *Sir Alfred McAlpine & Son (Northern) Ltd* (1985) under the ICE Fifth edition.

In that case the contractor was required to submit with his tender for works at Grimwith Reservoir a method statement showing that he would work upstream in constructing an outlet tunnel. This method statement became listed as one of the contract documents. In the event the contractor claimed it was impossible to work upstream and worked downstream. He then claimed he was entitled to a variation under clause 51 and payment accordingly. Finding in favour of the contractor Mr Justice Skinner said:

'In my judgment, the standard conditions recognise a clear distinction between obligations specified in the contract in detail, which both parties can take into account in agreeing a price, and those which are general and which do not have to be specified pre-contractually.

In this case the applicants could have left the programme and methods as the sole responsibility of the [contractor] under clause 14(1) and clause 14(3).

The risks inherent in such a programme or method would then have been the [contractor's] throughout.'

Changes from the ICE Sixth edition

Clause 14 of the D and C Conditions follows closely the wording of the ICE Sixth edition up to and including clause 14(6). Missing from clause 14 in the D and C Conditions are the last three sub-clauses of the ICE Sixth edition version of clause 14:

clause 14(7) – engineer's consent to the contractor's methods
clause 14(8) – delay and cost
clause 14(9) – responsibility unaffected by acceptance or consent.

However the provisions of these sub-clauses are not omitted entirely from the D and C Conditions. Clause 6 of the Conditions covers much of the content of clause 14(7) and clause 14(8) of the ICE Sixth edition; and clause 1(7) covers in full detail the content of clause 14(9).

Clause 14(1) – submission of programme

Clause 14(1)(a) of the Conditions requires the contractor to submit within 21 days of the award of the contract a programme 'showing the order in which he proposes to carry out the Works having regard to the provisions of Clause 42(1)'.

Taken literally, this might suggest that the only matters of note in the programme should be possession requirements but this is clearly not the intention.

As with other ICE forms of contract there is no specific requirement for a timescale in the programme – although it is clearly implied elsewhere. Clause 14(1) expressly requires only 'the order' of the works; and no doubt many employer's representatives will continue the practice, used with some wisdom by engineers under the ICE Fifth and Sixth editions, of accepting only 'the order'.

Failure to submit a programme

As with the ICE Sixth edition there are no express sanctions on the contractor if he fails to comply with the submission requirements of clause 14.

Two contractual remedies are available to the employer but both are extreme; the employer's representative could order a suspension under clause 40 and refuse the contractor's costs on the grounds the programme was 'necessary for the proper execution or for the safety of the Works'; or the employer's representative could activate the determination procedures of clause 65 on the grounds the contractor was persistently in breach of his obligations under the contract.

The powerful provision of the New Engineering Contract that half the value due on interim payments can be retained until the contractor has submitted an acceptable programme is not adopted.

General description and methods of construction

Clause 14(1)(b) requires the contractor to provide in writing, at the same time as he submits his programme, a general description of arrangements and methods of construction he proposes to adopt for carrying out the works.

Given the timescale for this requirement – within 21 days of the award of the contract – the level of information should not be expected to be any more than 'general'.

Rejection by the employer's representative

Clause 14(1)(c) requires the contractor to resubmit a revised programme within 21 days of any rejection by the employer's representative.

Note that clause 14(1)(c) applies only to the programme and not to arrangements and methods of construction.

Clause 14(2) – action by the employer's representative

Clause 14(2) requires the employer's representative to respond within 21 days of receipt of the contractor's programme by either:

- accepting the programme in writing
- rejecting the programme in writing with reasons
- requesting further information.

If the employer's representative does none of the above within 21 days, the programme is deemed to have been accepted as submitted.

Further information

In requesting further information on the programme the employer's representative may be seeking:

- clarification
- substantiation
- satisfaction on reasonableness having regard to the contractor's obligations.

Clause 14(3) – provision of further information

Clause 14(3) requires the contractor to respond within 21 days to the employer's representative's request for further information and the employer's representative in turn to respond within a further 21 days of its submission.

Clause 14(3) also provides that, if the contractor fails to provide further information within the 21 day timescale or such other period as allowed by the employer's representative, the programme shall be deemed to be rejected.

Clause 14(4) – revision of programme

Clause 14(4) entitles the employer's representative to seek a revised programme showing such modification as is necessary to ensure completion on time if it appears that the contractor's actual progress does not conform with the accepted programme.

It is doubtful if the employer's representative can use this clause when the contractor's progress does not conform but completion on time is not in doubt. The only modifications that can be required are those 'necessary' to ensure completion on time.

Clause 14(5) – design criteria

Clause 14(5) states that the employer's representative shall provide to the contractor such design criteria relevant to the employer's requirements to enable the contractor to comply with clause 14(6). That is, provide the design criteria which will enable the contractor to show that his methods of construction will not create detrimental stresses and strains in the permanent works.

The presence of this clause in the D and C Conditions is questionable given the contractor's responsibilities for design – not least those under clause 8(2)(b) where the contractor takes responsibility for the employer's design.

In the ICE Sixth edition the corresponding provision is understandable, but even there it is open to the criticism that the criteria described should be provided pre-tender if it is not to leave the employer liable to claims.

Clause 14(6) – methods of construction

Clause 14(6) gives the employer's representative a discretionary power to request information on methods of construction. The wording of the clause makes clear that it is concerned with detail likely to be beyond that submitted by the contractor originally under clause 14(1).

The purpose of the provision is primarily to enable the employer's representative to assess the effects of the contractor's methods on the permanent works as built and again the question must be asked how appropriate this is in a design and construct contract. Much would seem to depend on whether the employer intends that his employer's representative should become involved in detailed checks of the contractor's design.

8.11 Clauses 15 and 16 – contractor's superintendence and removal of employees

Clause 15(1) requires the contractor to provide superintendence during the construction and completion of the works, and as long thereafter as the employer's representative considers necessary.

Contractor's representative

Clause 15(2)(a) sets out the functions and authority of the contractor's representative. However, although it is clearly implied that such a representative will be appointed, there is no express provision to that effect as there is with the employer's representative.

The primary contractual functions of the contractor's representative are:

- to exercise overall superintendence of the works, and
- to receive on behalf of the contractor all consents, approvals etc given by the employer's representative.

Although there is nothing exceptional about the functions of the contractor's representative, his authority is certainly exceptional. Clause 15(2)(a) states that he has the authority to act and commit the contractor as if he were the contractor.

For further comment on this and the powers of the contractor's representative to delegate see Chapter 9.

Contractor's agent

Clause 15(3) requires the contractor to keep a competent and authorised agent on the works giving his whole time to superintendence of the same. The role of the agent is less comprehensive than that in the ICE Sixth edition. It is expressly stated to include duties to:

- superintend
- receive instructions on behalf of the contractor
- be responsible for the safety of all operations
- receive consents, approvals etc on behalf of the contractor from assistants of the employer's representative
- receive consents, approvals etc on behalf of the contractor's representative from the employer's representative.

Clause 16 – removal of contractor's employees

Clause 16 places an obligation on the contractor to employ only careful, skilled and experienced persons and entitles the employer's representative to require the removal of any person who is incompetent or negligent, guilty of misconduct or incautious of health and safety matters.

This is additional to the powers of removal of the employer's representative under clause 4(5) which apply to sub-contractors.

8.12 Clause 18 – boreholes and exploratory excavation

Clause 18 expands on the provisions in the ICE Sixth edition for making boreholes and exploratory excavations to take account of the contractor's design responsibilities.

Permission required

Clause 18(1) provides that if the contractor considers it necessary or desirable to make boreholes or carry out exploratory excavations, or investigations, he shall apply to the employer's representative for permission to do so. Such permission shall not be unreasonably withheld.

It may seem odd that the contractor, with his very broad design responsibility, should need permission to carry out tasks which, if not undertaken, might amount to negligence.

The intention of the provision is perhaps not so much to control what the contractor does in itself, but to control what the contractor does by way of damage to the employer's land and property.

Records and results

Clause 18(1) also provides that the contractor shall:

- comply with any conditions imposed by the employer's representative
- furnish copies of all information, records and test results
- furnish any expert opinion connected therewith.

Cost of making boreholes etc.

Clause 18(2) states that the cost of making boreholes etc, and of reinstatement – shall be borne by the contractor unless:

- they are a necessary consequence of a clause 12 situation, or
- they are a necessary consequence of a variation ordered under clause 51.

This is not dissimilar from the position in the ICE Sixth edition where most boreholes etc ordered by the engineer are related either to clause 12 or to clause 51 and the contractor gets paid accordingly.

Orders of the employer's representative

Clause 18(3) of the Conditions which states that any boreholes etc ordered by the employer's representative shall be deemed to be a variation under clause 51 unless covered by a contingency or prime cost item is no more than a repeat of the entire clause 18 of the ICE Sixth edition.

8.13 Clauses not used – clauses 13 and 17

Clause 13

In the ICE Sixth edition clause 13 fulfils a number of functions:

- It requires the contractor to construct and complete the works to the satisfaction of the engineer, save in so far as it is legally or physically impossible.
- In combination with clause 51 it requires the engineer to give instructions on overcoming legal or physical impossibility in so far as that can be achieved by the issue of variations.
- It requires the contractor to comply with the instructions of the engineer on any matter connected with the contract.
- It emphasises that the contractor must take instructions only from the engineer or the engineer's representative.
- It requires that materials, plant and labour and the speed of construction be acceptable to the engineer.
- It provides for the contractor to be paid reasonable extra cost for instructions not foreseeable at the time of tender and to be awarded an extension of time for any delay.

These are, by any standards, very important provisions and it is most unlikely that the draftsmen of the D and C Conditions chose to mark clause 13 – 'not used' – without serious consideration of the implications of the omission.

Impossibility

The provisions on impossibility have in part been moved to clause 8(1) but the D and C Conditions are by no means as clear as the ICE Sixth edition on how, if at all, impossibility short of frustration is to be tackled.

Instructions

The provisions on instructions are not obviously relocated in the D and C Conditions and indeed the Conditions are silent on the general power of the employer's representative to give instructions or the obligation of the contractor to comply with any such instructions. Clause 2 certainly implies that the employer's representative has the power to give instructions and that the contractor will comply – but it is only in respect of oral instructions that the contractor is expressly required to comply.

It may be said that the contractor's design responsibility eliminates the need for a general power to instruct and that, with a few limited exceptions, most instructions will amount to changes in the employer's requirements leading to variations under clause 51. However the weakness of this approach lies in clause 51 itself which gives the employer's representative power to issue variations only after 'consultation' with the contractor. This is not a solid a substitute for a power to instruct.

Clause 17 – setting-out

The omission of clause 17 from the D and C Conditions presents less difficulty than the omission of clause 13.

In the ICE Sixth edition clause 17 does little more than confirm the obvious, namely:

- the contractor is responsible for setting-out
- the contractor is paid only for rectifying errors arising from data supplied by the engineer

- the checking of setting-out by the engineer does not relieve the contractor of his responsibility.

These provisions are covered in various clauses of the D and C Conditions; although not with express reference to setting-out. But in any event there should be little need for the employer's representative to supply setting-out data or for him to get involved in checking setting-out.

8.14 *Schedule of express obligations and responsibilities of the contractor*

Clause	Description
4(2)(a)	to obtain the employer's consent prior to making any change in the named designer
4(4)	to remain liable for all work sub-contracted
6(1)(a)	to give notice in writing of any further information required
6(2)(a)	to submit designs and drawings to show the works will comply with the employer's requirements
6(2)(b)	to re-submit designs or drawings which in the opinion of the employer's representative do not comply with the employer's requirements or any other provisions of the contract
6(2)(c)	to give notice before modifying any design to which consent has been given
6(3)	to supply two copies of designs, drawings, specifications etc
7(2)	to keep one copy of documents, drawings etc on site
8(1)(a)	to design, construct and complete the works save in so far as it is legally or physically impossible
8(1)(b)	to provide all design services, labour, materials etc required for the works
8(2)(a)	to exercise all reasonable skill, care and diligence in carrying out design obligations

Clause	Description
8(2)(b)	to check the employer's design and accept responsibility therefor
8(3)	to the extent required by the contract to institute a quality assurance system
8(4)	to arrange and pay for statutory checks on design
8(5)	to take responsibility for the safety of the design and for the adequacy, stability and safety of all site operations and methods of construction
9	to enter into and execute a contract agreement when called on to do so
10(1)	to provide security for the performance of the contract if the contract so requires
11(2)	to inspect the site and to have satisfied himself before the award of the contract on the nature of the ground etc, risks, contingencies and other circumstances (deemed obligation)
11(3)(a)	to base his tender on information made available and own findings (deemed obligation)
11(3)(b)	to satisfy himself before the award of the contract as to the correctness and sufficiency of rates and prices (deemed obligation)
12(1)	to give written notice if unforeseen physical conditions encountered
12(2)	to give written notice if a claim is to be made for unforeseen physical conditions
12(3)	to give details of effects of, and proposals for dealing with, unforeseen physical conditions
14(1)(a)	to submit a programme within 21 days of award of the contract
14(1)(b)	to submit a general description of arrangements and methods of construction within 21 days of the award of the contract
14(1)(c)	to submit a revised programme within 21 days of any rejection
14(3)	to submit further information on the programme within 21 days of any request

Clause	Description
14(4)	to produce a revised programme when required
14(6)	to submit further details pertaining to the design and methods of construction if so requested
15(1)	to provide all necessary superintendence
15(3)	to appoint a full time contractor's agent
16(1)	to employ only careful, skilled and experienced staff and workmen
18(1)	to apply for permission to make boreholes or investigations etc considered necessary or desirable
19(1)	to have full regard for safety and security
20(1)	to take full responsibility for care of the works
20(3)	to rectify loss or damage to the works
21(1)	to insure the works in the joint names of the contractor and the employer
22(1)	to indemnify the employer against damage to persons and property
23(1)	to insure in the joint names of the contractor and the employer against third party claims
24	to indemnify the employer against claims for accidents or injury to workpeople
25(1)	to provide evidence of insurances prior to the commencement date
25(4)	to comply with conditions of insurance policies
26(1)	to give all notices and pay all fees required by statute etc
26(3)	to conform with statutes etc and to indemnify the employer against claims
28(1)	to indemnify the employer against claims in respect of patents etc
28(3)	to pay all tonnage etc for materials for the works
29(1)	to carry out the works without interference with traffic and adjoining properties and to indemnify the employer against claims

Clause	Description
29(2)	to carry out the work without unreasonable noise, disturbance or pollution
29(3)	to indemnify the employer against claims for noise, disturbance or pollution
30(1)	to use reasonable means to avoid damage to highways
30(2)	to be responsible for and to pay for strengthening bridges etc for the movement of contractor's equipment, temporary works or materials and to indemnify the employer against claims
31(1)	to afford reasonable facilities for other contractors
32	to give notice of the discovery of fossils, antiquities etc
33(1)	to clear the site on completion
35(1)	to provide returns of labour and contractor's equipment if the contract so provides
35(2)	to provide returns of labour and contractor's equipment when so requested
36(1)	to design, construct and complete the works in accordance with the contract
36(2)	to use materials and workmanship of the kinds described in the contract or where not so described as is appropriate in all the circumstances
36(3)(a)	to submit proposals for checking the design and setting-out of the works
36(3)(b)	to carry out checks and tests
36(4)	to provide assistance for measuring and testing
36(5)	to supply samples
36(6)	to consider the need for tests following variations
36(7)	to bear the costs of tests
37	to provide access to the works
38(1)	to afford opportunity for examination and measuring work about to be covered up and to give notice before covering up
38(2)	to uncover work and make openings as so directed

Clause	Description
39	to remove unsatisfactory work and materials when so instructed
40(1)	to suspend progress when so instructed
41(2)	to start the works as soon as practicable after the commencement date and to proceed with due expedition and without delay
42(4)	to bear all costs and charges for any additional access
44(1)	to provide details of any extension considered due
46(1)	to take steps as necessary to expedite progress
47(1)(b)	to pay liquidated damages for delay
48(1)	to undertake to finish outstanding works
48(3)	to provide operation and maintenance instructions prior to substantial completion
49(1)	to complete outstanding works by the times agreed or as soon as practicable during the defects correction period
49(2)	to deliver up the works at the end of the defects correction period in the condition required by the contract
49(3)	to bear the costs of work of repair due to own failures
50	to carry out searches etc to determine the cause of defects when so required
52(1)	to submit quotations for varied works
52(3)	to undertake daywork when so instructed
53(1)	to give notice in writing of claims and to keep contemporary records
53(2)	to permit inspection of records
53(3)	to submit a first interim account of claims and further accounts as requested
54(1)	to seek consent to remove contractor's equipment
54(2)	to transfer property in goods and materials if so instructed

Clause	Description
54(3)	to provide evidence of the vesting of goods and materials in the employer
54(4)(b)	to be responsible for loss or damage to goods and materials vested in the employer
54(5)	to bring to the attention of sub-contractors etc the title of the employer to goods and materials
54(6)	to deliver vested goods and materials to the employer upon cessation
54(7)	to incorporate provisions on vesting in sub-contracts
56(1)	to give notice if it is necessary to measure
56(3)	to furnish records for dayworks
58(1)	to obtain consent prior to commencing work on contingencies or prime cost items
60(1)	to submit interim statements as the contract prescribes
60(4)	to submit a final account with all supporting documentation not later than 3 months after the date of the defects correction certificate
61(3)	to submit operation and maintenance manuals prior to the issue of the defects correction certificate
63(3)	to remove contractor's equipment from site on abandonment
68(2)	to serve all notices on the employer at its last known address or registered office
69(2)	to allow the employer any decrease in cost in labour tax matters.

8.15 *Schedule of express obligations and responsibilities of the employer*

Clause	Description
2(2)	to notify the contractor in writing the name of the employer's representative
5(1)(c)(ii)	to pay the contractor's costs arising from ambiguities or discrepancies in the employer's requirements
5(2)	to assemble two complete copies of the contract upon the award of the contract
6(1)(b)	to pay the contractor's costs arising from failure to provide further information within a reasonable time
6(2)(d)	to pay the contractor's costs arising from failure to notify consent or withholding thereof in respect of designs and drawings
8(3)	to pay the contractor's costs arising from delay in approval to the contractor's quality assurance system
11(1)	to make available all information on ground conditions etc before tenders are submitted (deemed obligation)
12(6)	to pay the contractor's costs arising from unforeseen physical conditions
19(2)	to have full regard for safety on site in respect of own workpeople and to require other sub-contractors to so conform
20(1)(a)	to accept responsibility for care of the works after the issue of the certificate of completion for the whole of the works
20(1)(b)	to accept responsibility for care of parts or sections of the works after the issue of certificates of completion for the same
20(2)	to accept responsibility for the 'excepted risks'
22(2)	to accept responsibility for 'exceptions' to damage to persons and property
22(3)	to indemnify the contractor against claims in respect of the 'exceptions'
25(4)	to comply with the terms of insurance policies

Clause	Description
26(2)	to repay the contractor sums paid in respect of fees, rates and taxes
28(2)	to indemnify the contractor against claims for breach of patent rights etc arising from the employer's specifications, design or use of the works
29(4)	to indemnify the contractor against claims for noise, disturbance or pollution which is the unavoidable consequence of carrying out the works
30(3)(c)	to negotiate the settlement of claims and pay all sums due for damage to highways arising from the transport of materials
31(2)	to pay the contractor's costs arising from the provision of facilities for other contractors
32	to meet the expense of disposal of fossils etc
36(5)	to meet the cost of samples not clearly intended by or provided for in the contract
36(7)	to meet the costs of tests not clearly intended by or provided for in the contract and successful tests which are so intended or provided for
38(2)	to meet the costs of uncovering work and making openings of work found to be satisfactory
40(1)	to pay the contractor's costs arising from suspension of work
42(2)(a)	to give possession of so much of the site and access thereto as required to enable the contractor to commence and proceed
42(2)(b)	to give possession of further portions of the site as required by the contractor's clause 14 programme
42(3)	to pay the contractor's costs arising from failure to give possession
47(5)	to reimburse the contractor sums over-recovered as liquidated damages with interest
49(3)	to meet the cost of works of repair not due to contractor's failure of performance

Clause	Description
50	to meet the costs of searches for causes of defects when the contractor is not liable
51(3)	to pay a fair and reasonable value for all variations
56(3)	to pay for dayworks ordered by the employer's representative
58(2)	to pay for contingencies and prime cost items as valued
60(2)	to pay amounts due on interim certificates within 28 days of application
60(4)	to pay any balance due on the final account within 28 days of the payment certificate
60(6)	to release retention money as provided
60(7)	to pay interest on overdue payments
60(10)	to notify the contractor with full details when any amount paid is less than that certified
62	to meet the cost of urgent repairs not the contractor's liability
68(1)	to serve all notices on the contractor at the contractor's principal place of business or its registered address
69(2)	to meet the cost of any increases in labour tax matters
70(2)	to separately identify and pay VAT to the contractor.

The employer is additionally responsible for the performance of the employer's representative. And failure by the employer's representative to carry out his duties under the contract leaves the employer liable to the contractor for the consequences. See Chapter 9 for a schedule of the duties of the employer's representative and commentary thereon.

Chapter 9

Representatives of the parties

9.1 *Introduction*

The D and C Conditions envisage that both the employer and the contractor will appoint named representatives to facilitate the management of the contract. The probability is that on most projects other than those of considerable size, the representatives will be office based rather than site based and the Conditions allow for this by the appointment of assistants and a contractor's site agent.

The employer is obliged to appoint a representative by clause 2(2) and the Conditions cannot be operated without an employer's representative. There is no corresponding express obligation on the contractor to appoint his representative; the Conditions merely assume that the appointment is made. Failure by the contractor to appoint would probably be breach of an implied term but it would not be fatal to the contract.

Role of the representatives

The definitions in clause 1 of the Conditions require the representatives 'to act as such for the purposes of the contract'. 'Such' being the employer in one case and the contractor in the other. This gives the clearest possible indication that the representatives act principally in agency roles.

The matter is left in no doubt in respect of the contractor's representative by clause 15(2) which says:

> 'The Contractor's Representative shall have the authority to act and to commit the Contractor as if he were the Contractor for the purposes of the Contract.'

The agency role of the employer's representative is not emphasised with equal clarity and a number of clauses in the Conditions, notably clause 1(7)(a) on consents and approvals and clause 2(1)(b) 'no authority to amend the Contract nor to relieve the Contractor of any of his obligations' show that the employer's representative does not have full capacity to commit the employer.

The explanation for this lies in the different functions attributed to the two representatives. The employer's representative is responsible for administering the contract and is mentioned over 200 times in the Conditions. Much of his involvement is not in the nature of agency and it is not intended that the phrase 'Employer's Representative' should be synonymous with 'Employer'.

The contractor's representative on the other hand has no function other than to represent the contractor. Apart from clause 1 (definitions) and clause 15 (authority) the contractor's representative is mentioned only twice in the Conditions; in clause 51 (variations) and clause 56 (measurement).

Demise of the engineer

The departure in the D and C Conditions from the traditionally named 'engineer' as contract administrator follows the trend in many recently issued design and construct contracts. The IChemE Green Book has the 'project manager'; The Department of Transport form has the 'employer's agent'.

The logic of the change is that under a traditional civil engineering contract the engineer has four distinct functions:

- designer of the permanent works
- supervisor/inspector of construction
- contract administrator/certifier
- representative of the employer.

Under a design and construct contract the designer function is lost; the supervisor/inspector function is diminished and only the administrator/certifier and the employer's representative functions remain intact.

Thus it is said that since the lead role the engineer occupies under traditional contracting does not apply to design and construct contracting it is best to openly recognise the change by dropping the title.

In any event it has to be recognised that in design and construct contracting the employer has chosen a method of procurement where, once he has settled his requirements, representation rather than responsibility is what he intends thereafter. The title of 'employer's representative' is certainly more appropriate in this situation than 'engineer' with its wider associations.

Impartiality of the employer's representative

One of the less convincing reasons for changing from the engineer to the employer's representative is the suggestion put forward in some quarters that whereas the engineer is required to exercise and show impartiality in administering the contract no such requirement falls on a person entitled the employer's representative. This, it is said, permits the employer's representative to look after the employer's interests with a freedom denied to the engineer.

It would however be wholly wrong to assume that the need for impartiality can be discarded in this way. The requirement for impartiality in administering the contract remains as strong in the D and C Conditions as in any traditional ICE contract. Because of its importance the point is discussed in some detail later in this chapter.

9.2 *Role of the employer's representative*

The employer's representative has duties to fulfil under the D and C Conditions and obligations to the employer under the terms of his engagement. He may also have wider duties outside the contract.

Such duties could include:

- preliminary design
- obtaining planning approvals
- drafting the employer's requirements
- management of the tender process
- advising on design changes
- overall project management.

In some cases the employer's representative will undertake the whole range of duties; in other cases the various duties may be spread between in-house and external teams; and in other cases the division of duties may be along professional lines following the practice popular in the building industry for design and build projects.

However it should make no difference to the performance of the employer's representative under the contract whether or not he carries out solely that role or he has a wider brief. The contractor is entitled to assume that whoever carries out the duties of the employer's representative will do so in a professional manner and that he will issue his certificates with fairness and impartiality.

Impartiality

The D and C Conditions do not have a provision matching clause 8(2) of the ICE Sixth edition which states that the engineer is to act impartially within the terms of the contract. Some commentators have concluded from this that under the D and C Conditions the employer's representative need not be impartial.

The word 'impartially' can lead to some difficulty since in the narrow sense of affording both parties a hearing it can be distinguished from fairness as this extract from the judgment of Viscount Dilhorne in the case of *Sutcliffe* v. *Thackrah* (1974) illustrates:

> 'Here the architect is required to issue interim certificates for the "total value of the work properly executed" and in valuing that work he has to use his professional skill and knowledge. He is not employed to be unfair to the builder. He is not required to determine a dispute between his employer and the builder. As I see it, there is no question of his having to act impartially between them. He must, if he exercises his professional skill and knowledge as it should be exercised, assess the total value of the works properly executed at what he honestly believes to be its true value. If he does that, he is acting fairly.'

But the wider meaning of impartially can be gathered from the *Shorter Oxford Dictionary* which defines impartial as: 'not favouring one more than the other, unprejudiced, unbiased, fair, just, equitable'. And this is more in accord with the usual interpretation of the phrase 'duty to act impartially'.

Lord Reid in his judgment in the *Sutcliffe* v. *Thackrah* case expressed such a duty in these words:

> 'It has often been said, I think rightly, that the architect has two different types of function to perform. In many matters he is bound to act on his client's instructions, whether he agrees with

them or not; but in many other matters requiring professional skill he must form and act on his own opinion.

Many matters may arise in the course of the execution of a building contract where a decision has to be made which will affect the amount of money which the contractor gets. Under the RIBA contract many such decisions have to be made by the architect and the parties agree to accept his decisions. For example, he decides whether the contractor should be reimbursed for loss under clause 11 (variation), clause 24 (disturbance) or clause 34 (antiquities); whether he should be allowed extra time (clause 23); or where work ought reasonably have been completed (clause 22). And, perhaps most important, he has to decide whether work is defective. These decisions will be reflected in the amounts contained in certificates issued by the architect.

The building owner and the contractor make their contract on the understanding that in all such matters the architect will act in a fair and unbiased manner and it must therefore be implicit in the owner's contract with the architect that he shall not only exercise due care and skill but also reach such decisions fairly holding the balance between his client and the contractor.'

The application of those principles to civil engineering contracts was confirmed in the New Zealand case of *Canterbury Pipe Lines* v. *Christchurch Drainage Board* (1979) where it was said, referring to *Sutcliffe* v. *Thackrah* and other cases:

'In our opinion it should be held in the light of these authorities that in certifying or acting under Clause 13 here the engineer, though not bound to act judicially in the ordinary sense, was bound to act fairly and impartially.'

There can be little doubt then that notwithstanding the absence of an express provision for impartiality in the D and C Conditions the employer's representative has an implied duty to act fairly and impartially when certifying or carrying out other duties under the contract which require the exercise of his discretion.

Certifying

There is virtually no difference between the D and C Conditions and the ICE Sixth edition with regard to the issue of certificates.

The provisions for payment certificates, certificates of substantial completion and the defects correction certificate are nearly identical. So also are the provisions for extending time.

In all of these the role of the employer's representative in the D and C Conditions is the same as that of the engineer in the ICE Sixth edition.

Supervising

Surprisingly, perhaps, the Conditions themselves indicate little change in the role of the employer's representative with regard to supervision of the works from that of the engineer in the ICE Sixth edition.

Some commentators have suggested that the employer's representative has no supervisory role – a thought derived it seems from the use of the word 'watch' in the D and C Conditions instead of the corresponding words 'watch and supervise' in the ICE Sixth edition.

But the extent of the employer's representative supervisory role is determined by the various provisions of the Conditions not by reference to the occasional phrase and most of the provisions in the ICE Sixth edition which are of a supervisory nature are to be found in the D and C Conditions including:

clause 4 – power to order removal of sub-contractors
clause 12 – consent to contractor's measures
clause 14 – acceptance of programme
clause 16 – removal of contractor's employees
clause 18 – making of boreholes
clause 19 – instructions on safety and security
clause 36 – testing
clause 38 – examination of work before covering up
clause 39 – removal of unsatisfactory work and materials
clause 40 – suspension of work
clause 46 – rate of progress.

It may well be that the employer's representative is not intended in practice to have the same active involvement in supervising the works as an engineer. And certainly the provision in clause 8(3) for the contractor to institute 'to the extent so required' a quality assurance system supports this view.

But anyone taking on the position of employer's representative under the D and C Conditions needs to clear this matter with the employer. Failure to use the powers given by the Conditions could amount to breach of terms of engagement – particularly if the contractor's performance is allowed to fall below standard.

Decision making

One important difference between the employer's representative under the D and C Conditions and engineer under the ICE Sixth edition is that the employer's representative has no power to make decisions of finality.

The engineer's decision under clause 66 is omitted with the effect that any decision of the employer's representative remains open to challenge by either party and does not become settled after a specified period.

9.3 *Duties and authority of the employer's representative*

Clause 2(1)(a) – duties and authority

Clause 2(1)(a) of the D and C Conditions states that the employer's representative shall carry out the duties and may exercise the authority specified in or to be implied from the contract. The ICE Sixth edition has similar provisions.

As a general rule, duties in the Conditions, as in the most standard forms, are preceded by the word 'shall'. Powers, which indicate exercise of authority, or the use of discretion, are preceded by the word 'may'.

The schedule later in this chapter lists the various duties and powers of the employer's representative in the D and C Conditions.

The importance of clause 2(1)(a) which might seem to be no more than a statement of the obvious is that it makes failure by the employer's representative to carry out his duties a breach of contract for which the employer is financially responsible. Many such potential breaches are covered elsewhere in the Conditions with appropriate remedies for the contractor, for example, late supply of information under clause 6 or failure to certify under clause 60. But where there is a duty but no express remedy for default, for example, failure to give instructions on fossils under clause 32 or failure to grant interim extensions of time 'forthwith' under clause

44(3) the contractor will find clause 2(1)(a) a useful starting point for a claim. It will not be necessary to establish firstly an implied term that the employer's representative will carry out his duties. The express term applies.

Clause 2(1)(b) – no authority to amend

Clause 2(1)(b) states that the employer's representative has no authority to amend the contract nor to relieve the contractor of his obligations except as expressly stated in the contract.

This is another provision taken from the ICE Sixth edition but it may be more appropriate here because whereas it is difficult to see how the engineer can be assumed to have any authority other than from the contract, the employer's representative has a more apparent agency role and might be thought to have wider authority.

The effect of clause 2(1)(b) is to curtail the agency role of the employer's representative. The employer's representative cannot, for example, relieve the contractor of his obligation to pay liquidated damages for late completion. Clearly the employer can do so if he wishes but the contractor cannot rely on dispensations granted by the employer's representative unless evidence of additional authority is provided.

Personal liability of the employer's representative

The employer's representative will, under the usual rules of law, be liable to the employer under his contract of engagement for breaches of duty. For example, it was held in the case of *Sutcliffe* v. *Thackrah* mentioned earlier that an employer could recover his loss from his architect who had negligently over-certified prior to the contractor going into liquidation. The architect's defence that he was acting in an arbitral capacity when certifying was not accepted by the court.

Many legal commentators felt that a similar remedy would be open to a contractor as an action in negligence when the engineer had caused loss by under-certifying. But the Court of Appeal in *Pacific Associates* v. *Baxter* (1988) rejected such a claim holding that the engineer had no duty of care in respect of economic loss to the contractor and that the arbitration clause in the contract provided the contractor's remedy. In so far as this decision is applicable the employer's representative is effectively protected from direct claims by the contractor arising out of alleged failure to carry out duties or to exercise discretion fairly and impartially.

That, of course, is the position of the contractor and it is not to say that the employer could not recover from the employer's representative monies paid out by the employer to the contractor for the defaults of the employer's representative. Such recovery would be under the contract between the employer and his representative and it would not meet the obstacle of recovery of economic loss which exists in negligence actions in tort.

9.4 *Name of the employer's representative*

Clause 1(1)(c) defines the employer's representative as the person appointed by the employer. The name of the employer's representative is either to be stated in the appendix to the form of tender or notified to the contractor under clause 2(2) within seven days of the award of contract or before the commencement date.

Unlike the ICE Sixth edition which permits a firm to be appointed as the engineer, the D and C Conditions seem to preclude a firm being appointed as the employer's representative. This has some interesting implications for the professional indemnity of the named individual if he works within a firm.

Replacement of the employer's representative

Clause 1(1)(c) permits the employer to change his appointment 'from time to time' subject only to notifying the contractor in writing of the change and, under clause 2(2), the name of the replacement.

This freedom is not common to all design and construct forms and in MF/1, for example, the employer cannot replace the engineer without the contractor's prior consent.

No need to be chartered

The D and C Conditions do not require the employer's representative to be a chartered engineer in line with the ICE Sixth edition. Nevertheless the technical aspects of the employer's representative's duties imply that he will be more than a contract administrator and the interest being shown by some of the larger quantity surveying/project management firms in civil engineering contracts will be better directed towards the New Engineering Contract than to the D and C Conditions.

9.5 Delegation

Clause 2(3) permits the employer's representative to delegate his duties and authorities except those relating to decisions or certificates issued under:

clause 12(6) – unforeseen conditions
clause 15(2)(b) – delegation by contractor's representative
clause 44 – extensions of time
clause 46(3) – accelerated completion
clause 48 – substantial completion
clause 60(4) – final payment certificate
clause 61 – defects correction certificate
clause 65 – default of the contractor.

The delegation can be to any person, including assistants, and remains in force until revoked. The delegation is to be in writing and does not take effect until a copy is delivered to the contractor or his representative.

Division of duties

The wording of clause 2(3), in particular the reference to 'assistants' in the plural, suggests that the employer's representative can divide his duties and authorities by delegation to a number of persons. There is certainly a case for this given the variety of his duties but care needs to be taken to ensure that overlap does not lead to confusion.

Thus if the employer's representative delegates to various assistants each with individual responsibilities for design, construction and valuation all three could have an interest in the operation of clause 12(4) – measures arising from unforeseen conditions – but, of course, only one could have the delegated power.

Delegation not general

Although clause 2(3) does not expressly preclude general delegation the words 'any of his duties and authorities' suggest that notices of delegation should be specific. This is supported by the phrase 'scope of authority' in clause 2(4) relating to the duties of assistants.

Nature of delegation

The D and C Conditions do not resolve the debate on whether powers which are delegated reside solely with the person to whom they are delegated unless and until revoked, or whether they are shared with the original holder of the powers.

There are two views on this but practical considerations in construction contracts suggest that a power should not be exercisable simultaneously by different persons. However if the correct view is that delegated powers are shared powers this would not diminish the effectiveness of actions taken by one of the persons using his delegated power.

Authority to delegate

The D and C Conditions have nothing to say on restrictions on the authority of the employer's representative – unlike the ICE Sixth edition which requires such restrictions to be specified in the appendix to the form of tender. Consequently the contractor is entitled to assume that the employer's representative has the authority of the employer to delegate.

But in reality the employer, having selected an individual to be his representative, may not favour delegation to assistants particularly on such important matters as varying the employer's requirements under clause 51 or suspending the works under clause 40.

This is a matter solely between the employer and the employer's representative but it justifies careful consideration as a matter of policy before the contract commences.

9.6 Assistants

Clause 2(4)(a) permits the employer's representative to appoint any number of assistants 'for the carrying out of his duties under the Contract'. The contractor is to be notified of the names, duties and scope of authority of such assistants but there is no requirement in the clause itself for the notification to be in writing.

No engineer's representative equivalent

There is no provision in the D and C Conditions for the employer's representative to appoint a specific assistant equivalent to the engineer's representative in the ICE Sixth edition to act as the resident engineer.

Nevertheless, for most projects of any size employers are likely to retain some elements of the traditional site supervision team – at least until confidence and experience in the quality assurance systems operated by contractors makes such supervision superfluous. And where there is a team there needs to be a leader so the position of resident engineer and, indeed, the title will still have its place in many contracts.

Note, however, that there is no provision in the D and C Conditions for assistants to appoint assistants in like manner to the appointment of assistants by the engineer's representative in the ICE Sixth edition. The effect of this may be quite significant since it means there can be only one stage of delegation and sub-delegation as permitted by the ICE Sixth edition is not envisaged. Consequently if an assistant called the resident engineer is given delegated powers he cannot pass down those powers to his own assistant resident engineers or clerks of works. They can only acquire delegated powers directly from the employer's representative.

Duties of assistants

Unless assistants are given delegated powers they have no duties or authority under the D and C Conditions other than to watch and give instructions on safety and security under clause 19.

The distinction between the word 'watch' in the D and C Conditions and the corresponding phrase 'watch and supervise' in the ICE Sixth edition amounts to very little since under the ICE Sixth edition only the engineer is given active powers to supervise.

Clause 2(4)(b) which states that assistants may be appointed specifically to watch the construction of the works indicates that not all assistants need to be given delegated powers to have a role on the site. This would seem to go without saying but perhaps the provision is of some significance in connection with clause 37 (access to site) in allowing persons authorised by the employer's representative to have access at all times.

No authority to relieve the contractor

Clause 2(4)(c) states that assistants have no authority to relieve the contractor of his duties or obligations nor have they authority, unless exercising delegated powers, to:

- order work involving delay or extra payment, or
- make any variation of or in the works.

The phrase 'any variation of or in the Works' is interesting since under clause 51 the employer's representative is only empowered to vary the employer's requirements. The 'Works' by definition includes both the permanent works and the temporary works so the phrase may be used to cover instructions relating to temporary works as well as to changes in the employer's requirements.

9.7 *Instructions*

To be in writing

Clause 2(5)(a) requires all instructions given by the employer's representative or any person exercising delegated authority to be in writing. However, in common with the policy of many other contracts, the clause also requires the contractor to comply with instructions given orally.

With the omission of clause 13 from the D and C Conditions this latter point is worth noting as it is the only express requirement in the Conditions that the contractor must comply with instructions. However it does not address the question of an instruction for which the contractor thinks there is no authority. For further comment on this see below under 'specification of authority'.

Confirmation of oral instructions

Clause 2(5)(b) states that any oral instruction shall be confirmed in writing 'as soon as is possible under the circumstances'.

The clause then goes on to say that if the contractor confirms an oral instruction in writing and it is not contradicted in writing 'forthwith' then it is deemed to be an instruction in writing.

The *Shorter Oxford Dictionary* gives the meaning of 'forthwith' as: immediately, at once, without delay or interval. This suggests that a response must be within days rather than within weeks.

Specification of authority

Clause 2(5)(c) follows a provision new to the ICE Sixth edition but long overdue to ICE contracts generally requiring the employer's representative or any person exercising delegated authority to specify in writing, if the contractor so requests, under which of his duties or authorities any instruction is given.

This provision added to that for instructions to be in writing should be a powerful incentive for the employer's representative and his assistants to think carefully before acting on any matter.

In the event that the employer's representative is unable to show any contractual authority for an instruction given by himself, or one of his assistants (in response to a reference on dissatisfaction under clause 2(6)), the proper course of action would be to withdraw the instruction.

This indicates that if the contractor is in any doubt on the propriety of an instruction he should raise the matter under clause 2(5)(c) before complying. The contractor will find no entitlement in the Conditions for complying with an instruction given without authority.

Reference on dissatisfaction

Under clause 2(6) if the contractor is dissatisfied with any act, instruction or decision of any assistant he is entitled to refer the matter to the employer's representative for a decision. This would appear to apply to both the express power of 'watching' and to delegated powers.

Commonsense again suggests that the contractor should raise the matter before complying with any instruction etc but this may not be the full intention of the clause. Clause 66 on the settlement of disputes says that no notice of dispute shall be served until the procedures available elsewhere in the contract have been used.

Clause 2(6) is one of the few available procedures so it would seem to apply to disputes both before and after compliance.

There is an argument that the contractor should lose his right to reimbursement for complying with an instruction from an assistant if he objects but fails to raise the matter with the employer's representative before complying. But clause 2(6) says the contractor 'shall be entitled' to refer the matter, it does not say that he is obliged to do so. The risk that an assistant with delegated powers might give erroneous instructions remains with the employer.

9.8 *Express duties and powers of the employer's representative*

The following schedules show the express duties and powers of the employer's representative under the D and C Conditions compared with the corresponding duties and powers of the engineer under the ICE Sixth edition.

Express duties of the employer's representative

Clause no.	Duty	Sixth edition equivalent
2(1)(a)	to carry out the duties specified in or to be implied from the contract	2(1)(a)
2(5)(b)	to confirm oral instructions in writing as soon as possible	2(6)(b)
2(5)(c)	to specify in writing under which duty or authority an instruction is given	2(6)(c)
5(1)(c)(i)	to explain or adjust ambiguities or discrepancies within the employer's requirements	5
5(1)(c)(ii)	to consider delay and cost arising from the above	13(3)

Clause no.	Duty	Sixth edition equivalent
6(2)(b)	to inform the contractor if designs or drawings do not comply with the employer's requirements	—
6(2)(d)	to consider delay and cost arising from failure to notify the contractor within a reasonable period of consent or withholding of consent to designs or drawings	—
8(3)	to consider delay and cost arising from failure to notify the contractor within a reasonable period of consent or withholding of consent to the contractor's quality assurance system	—
12(5)	to inform the contractor if physical conditions could reasonably have been foreseen	12(5)
12(6)	to consider delay and cost arising from unforeseen physical conditions	12(6)
14(2)	to accept, reject or request further details of the contractor's programme	14(2)
14(3)	to accept or reject the contractor's programme following receipt of further information	14(3)
14(5)	to provide design criteria relevant to the employer's requirements	14(5)
26(3)(b)	to issue instructions if the employer's requirements preclude conformity with statutes	26(3)(b)

Clause no.	Duty	Sixth edition equivalent
31(2)	to consider delay and cost arising from provision of facilities for other contractors	31(2)
38(1)	to attend for the purposes of examining or measuring foundations	38(1)
40(1)	to consider delay arising from suspension of the works	40(1)
42(3)	to consider delay and cost arising from failure to give possession and to notify the contractor with copy to the employer	42(3)
44(2)	to make an assessment of delay upon receipt of an application for an extension of time	44(2)
44(3)	to grant interim extensions of time forthwith	44(3)
44(4)	to assess extensions of time at the due date for completion	44(4)
44(5)	to review extensions of time at substantial completion	44(5)
46(1)	to notify the contractor if the rate of progress is too slow	46(1)
47(6)	to inform the contractor and employer if there is unavoidable delay after liquidated damages have become payable	47(6)
48(2)	to issue a certificate of substantial completion or give instructions of work to be done upon receipt of notice from the contractor	48(2)

Clause no.	Duty	Sixth edition equivalent
48(4)	to issue a certificate of substantial completion for any substantial part of the works occupied or used by the employer	48(3)
52(2)(b)	to accept or negotiate with the contractor on the valuation of variations	—
52(2)(d)	to notify the contractor of a fair and reasonable valuation in the absence of agreement	52(1)
56(1)	to give reasonable notice if the contractor is required to attend for measurement	56(3)
60(2)	to certify interim payments within 28 days of the contractor's application	60(2)
60(4)	to certify the amount finally due within three months of receipt of the contractor's final account	60(4)
61(1)	to issue a defects correction certificate stating the date on which the contractor has completed his obligations	61(1)
62	to inform the contractor if urgent repairs are needed	62
64(3)	to consider delay and cost arising from suspending or reducing the rate of work in consequence of failures in payment	—
65(4)	to carry out a valuation at the date of termination due to contractor's default	63(5)

Express powers of the employer's representative

Clause No.	Power	Sixth Edition Equivalent
1(1)(s)	to agree 'other places' as forming part of the site	1(1)(v)
2(1)(a)	to exercise authority specified or to be implied from the contract	2(1)(a)
2(3)	to delegate duties and authorities to any person	2(4)
2(4)	to appoint any number of assistants	2(5)
4(5)	to require the contractor to remove any sub-contractor for misconduct, incompetence, negligence, lack of regard for safety	4(5)
12(4)(a)	to give consent to the contractor's measures for dealing with unforeseen conditions	12(4)(b)
12(4)(b)	to order a suspension or issue a variation for unforeseen conditions	12(4)(d)
14(3)	to allow the contractor an additional period within which to supply additional information	14(3)
14(4)	to require the contractor to produce a revised programme	14(4)
15(1)	to require the contractor to provide superintendence after completion	15(1)
16(1)	to require the contractor to remove any employee for misconduct, incompetence, negligence or lack of regard for safety	16(1)

Clause no.	Duty	Sixth edition equivalent
18(3)	to order the contractor to make boreholes or carry out exploratory excavation	18
19(1)	to give instructions on safety and security	19(1)
20(3)(b)	to give instructions to rectify loss or damage	20(3)(b)
30(3)(c)	to assess contractor's responsibility for damage to highways	30(3)
31(1)	to require the contractor to afford reasonable facilities for other contractors	31(1)
32	to give orders on the examination and disposal of fossils and antiquities	32
33(2)	to allow a reasonable time for disposal of the contractor's equipment	53(3)
35(1)	to prescribe the form and intervals for returns of the contractor's plant and labour	35
35(2)	to require delivery of plant and labour returns where the contract contains no provision for such	—
36(3)	to require further tests on workmanship and materials	36(1)
36(4)	to require samples for testing	36(1)
38(2)	to give directions on uncovering any part of the works or making openings	38(2)
39(a)	to give instructions on the removal of materials which do not comply with the contract	39(1)(a)

Clause no.	Duty	Sixth edition equivalent
39(b)	to give instructions on replacement with materials which do comply with the contract	39(1)(b)
39(c)	to order the removal and proper replacement of materials and workmanship which do not comply with the contract	39(1)(c)
40(1)	to order suspension of progress of the works	40(1)
44(2)(b)	to make an assessment of delay in the absence of a claim	44(2)(b)
46(1)	to consent to steps to expedite progress	46(1)
46(2)	to give permission to work at night or on Sundays	46(2)
46(3)	to request accelerated completion	46(3)
48(5)	to issue a certificate of substantial completion in respect of any part of the works which has been substantially completed	48(4)
49(1)	to agree the time for the completion of outstanding work	49(1)
50	to require the contractor to carry out searches and trials to determine the cause of any defect or fault	50
51(1)	to vary the employer's requirements after consultation with the contractor	—

Clause no.	Duty	Sixth edition equivalent
52(1)	to request the contractor to submit a quotation for varied work	—
52(3)	to order additional or substituted work to be executed on a daywork basis	52(3)
53(2)	to instruct the contractor to keep contemporary records for claims	52(4)(c)
53(3)	to require the contractor to send up to date accounts of claims with accumulated totals	52(4)(d)
54(1)	to give consent to removal of contractor's equipment	53(1)
54(2)	to direct the transfer of property in goods and materials to the employer	54(1)
58(1)	to give consent to the commencement of work on contingency or prime cost items	58(1) & (2)
60(4)	to require information for the verification of the final account	60(4)
60(8)	to omit from any certificate the value of work done or goods or materials supplied or services rendered in the event of dissatisfaction and to modify previous sums and statements	60(8)
65(1)	to certify in writing the defaults of the contractor	63(1)(b)
65(3)	to instruct the contractor to assign the benefit of agreements to the employer.	63(3)

9.9 *The contractor's representative*

Definition

The contractor's representative is defined in clause 1(1)(d) as the person appointed by the contractor:

- to act as such for the purposes of the contract, and
- notified in writing as such to the employer.

By clause 1(1)(d) the contractor is permitted to appoint some other person from time to time providing the employer is again notified in writing.

Appointment

There is no express requirement in the Conditions that the contractor shall appoint a representative but it is certainly implied. The omission of an express provision may be a drafting oversight which could usefully be corrected.

Indeed many employers may wish to go further and follow the policy of the New Engineering Contract in requiring the contractor's key personnel, including the contractor's representative, to be named in the submission; and not permitting replacement without consent; and then only by persons with similar qualifications and experience.

Duties and authority

Clause 15(2) deals briefly with the duties and authority of the contractor's representative.

Clause 15(2)(a) states that the contractor's representative shall exercise overall superintendence 'of the Works' on behalf of the contractor.

'The Works' as used here includes design so clearly the contractor's representative is not to be seen only as a high-powered site manager. This is confirmed by the final sentence in clause 15(2)(a) which reads:

> 'The Contractor's Representative shall have the authority to act and to commit the Contractor as if he were the Contractor for the purposes of the Contract.'

This is likely to have most effect where there is provision in the contract for agreements to be reached between the contractor and employer. For example, in connection with acceleration under clause 46(3); the agreement to variations under clause 51 or the valuation of variations under clause 52.

Some contractors may deal with this by appointing only directors or senior managers to act as the named contractor's representative but whoever is appointed needs to have a hands-on role in the running of the contract.

Receipt of consents, approvals etc.

Clause 15(2)(a) also states, in addition to the overall superintendence point, that the contractor's representative shall receive on behalf of the contractor all:

- consents
- approvals
- orders
- instructions
- information

given by the employer's representative.

The list, it will be observed, does not include 'notices'. Perhaps this is because clause 68(1) requires all notices to be served at the contractor's principal place of business or, in the event of the contractor being a company, at its registered office. Fortunately, for the practical running of the contract, having regard to the required destination of the notices, the occasions when the employer's representative has to give notice to the contractor under the Conditions are few. Clause 46(1) – rate of progress is one example. Clause 60(10) – payment advice is another.

The employer's representative does have to give written notice to the contractor's representative under clause 56(1) on measurement (when that clause applies) but that is obviously distinguishable from a notice to the contractor.

Delegation

Clause 15(2)(b) permits the contractor's representative to delegate any of his responsibilities to a nominated deputy with the prior written agreement of the employer's representative.

Information and decisions from such a nominated deputy are then treated as though from the contractor's representative himself.

Note, however, that the Conditions do not give the contractor's representative responsibilities other than in:

- clause 15(2)(a) – superintendence and receipt of consents, approvals etc
- clause 15(3) – where the contractor's representative is also the contractor's site agent
- clause 56(2) – attending for measurement where this is appropriate.

Chapter 10

Assignment and sub-contracting

10.1 *Introduction*

The provisions for assignment and sub-contracting in the D and C Conditions are broadly the same as in the ICE Sixth edition. Thus neither party may assign the contract or any part without the consent of the other party but the contractor may sub-contract any part of the construction of the works without first obtaining consent. This latter freedom it must be said does not extend to changing the designer named in the appendix to the form of tender without the employer's consent.

No provision for nominated sub-contractors

The absence of any provision for nominated sub-contractors in the Conditions is, however, a major departure from traditional ICE conditions. But this new approach is in keeping with the policy of most other standard design and construct forms, including JCT 81. The reasons are obvious enough in principle. The contractor should be given maximum freedom to meet the employer's requirements by his own choice; and the potential pitfalls of nomination, which arguably outweigh the benefits even in traditional forms, are infinitely greater in design and construct. In practice, however, nomination is likely to live on, albeit under a different name and different contractual arrangements. The Conditions do not actually provide for the naming of sub-contractors in the employer's requirements, as does JCT 81, but if the employer undertakes detailed design himself the practice of naming sub-contractors may well be unavoidable.

The designer as a sub-contractor

Consulting engineering firms acting in the capacity of designer under the Conditions may not be too happy to find themselves regarded as sub-contractors. For some, the thought of working for contractors rather than supervising them will not be welcome. These difficulties will be resolved in time by changes in attitudes and, of course, no one expects to find consultants engaged under the FCEC blue form of sub-contract. Nevertheless there may be contractual implications in treating designers as sub-contractors even if such implications are not immediately apparent.

Sub-contracting in design and construct contracts

With the exception that a named designer can only be changed with the employer's consent, the Conditions impose no particular restraints on sub-contracting. There is no recognition of the special obligations of design and construct contracts. However, it can be argued that the employer has a greater interest in sub-contracting arrangements under such contracts than under traditional forms for a variety of reasons. Completion in the event of the main contractor's default is an obvious example – the employer might then want to retain or somehow re-engage key sub-contractors to ensure completion. To cover this, some design and construct forms require sub-contracting to be under approved terms; or under a standard form of sub-contract; or at least to incorporate certain provisions on a back to back basis with the main contract. The D and C Conditions do none of these except in the limited requirement of clause 54(7) relating to payment of goods and materials before delivery to site – and this is simply a repeat of a long standing provision in traditional ICE forms.

10.2 Assignment and sub-contracting generally

Clause 3 of the Conditions deals with assignment and clause 4 with sub-contracting. The practical man will have no difficulty in recognising that the first is intended to deal with the transfer of legal rights and the second with the fulfilment of contractual obligations by the employment of others.

The legal distinction between the two, however, is by no means clear cut and Mr I N Wallace in his Commentary on the ICE Conditions of Contract, Fifth edition, made this point:

'In the present clauses the words "assign the contract" in clause 3

> and "sub-let the whole of the Works" in clause 4 are probably synonymous, or very nearly so, as meaning vicarious performance of the whole of the works.'

Mr Wallace explained this as follows:

> 'English law does not permit the assignment of contractual liabilities, so that a contractor could never by a purported assignment or sub-contract escape his personal contractual and financial responsibility for the work. The law does, however, in general permit *vicarious performance* of a contractual liability, and it is this rather than assignment in the legal sense at which the prohibitions in these clauses are presumably aimed'

The views of Mr Wallace and others were examined by Lord Justice Staughton in the cases of *Linden Gardens Trust Ltd* v. *Lenesta Sludge Disposals Ltd* and *St Martins Property Corporation Ltd* v. *Sir Robert McAlpine & Sons Ltd* (heard together 1992) when he said of the JCT 63 building contract:

> 'Faced with this impressive near consensus of opinion, I conclude that the words "assign this contract" in clause 17(2) do indeed refer to the delegation of the contractor's obligations by sub-contracting them all. They do not include assignment (in its legal sense) of the benefit of the contract or any part of it. Although there is less reason for the employer to wish to sub-contract his obligations, or for the contractor to prohibit him from doing so, in my judgement the words "assign this contract" in clause 17(1) must be given the same meaning. They prohibit vicarious performance by the employer without consent, and do not prohibit assignment by the employer of the benefit of the contract or any part of it.'

However, when the House of Lords came to hear these same cases in 1993 the above views were firmly rejected. Lord Browne-Wilkinson put the matter as follows:

> 'Like the majority of the Court of Appeal, I am unable to accept this argument. Although it is true that the phrase "assign this contract" is not strictly accurate, lawyers frequently use those words inaccurately to describe an assignment of the benefit of a contract since every lawyer knows that the burden of a contract cannot be assigned: see, for example, *Nokes* v. *Doncaster Amal-*

gamated Collieries Ltd [1940] AC 1014, 1019–1020. The prohibition in clause 17(2) against sub-letting "any portion of the Works" necessarily produces a prohibition against the sub-letting of the whole of the works: any sub-letting of the whole will necessarily include a sub–letting of a portion and is therefore prohibited. Therefore there is no ground for reading the words "assign this contract" in clause 17(1) as referring only to sub-letting the whole. Decisively, both clause 17(1) and (2) clearly distinguish between "assignment" and "sub-letting": it is therefore impossible to read the word "assign" as meaning "sub-let". Finally, I find it difficult to comprehend the concept of an employer "sub-letting" the performance of his contractual duties which consist primarily of providing access to the site and paying for the works.'

Before considering further why the law on assignment and sub-contracting is so complex it is first necessary to consider the doctrine of privity of contract.

Privity of contract

Privity of contract relates to those bound by the contract. It identifies the persons, or parties, the contract legally affects. The essence of the doctrine is that only the parties acquire rights or obligations under the contract. It was succinctly stated by Lord Haldane in the case of *Dunlop Pneumatic Tyre Co. Ltd* v. *Selfridge & Co. Ltd* (1915) as follows:

> 'In the law of England certain principles are fundamental. One is that only a person who is a party to a contract can sue on it.'

The doctrine has both negative and positive implications. A sub-contractor cannot sue the employer for work done but not paid for by the main contractor. Bad news for the sub–contractor, but good news for the employer. Conversely the employer cannot sue the sub–contractor for faulty work in contract; bad news for the employer, but good news for the sub–contractor.

The doctrine has long been seen as a source of potential injustice, particularly in cases involving the breaking of promises intended to confer benefits on third parties. *Tweddle* v. *Atkinson* (1861) and *Beswick* v. *Beswick* (1968) are two of the better known. But even in construction contracts the rigidity of the doctrine can be seen as imposing artificial restraint on commercial intentions. This passage is taken from Emden's Construction Law:

> 'To many observers, however, the idea that a promise which is reproduced uniformly and to the knowledge of all parties throughout a series of contracts can be enforced only between immediate contractual neighbours, and not among parties at one or more contractual removes, seems commercially (if not rationally) absurd.'

Chain of responsibility

In fact if a chain of contractual responsibility is established – employer to sub-contractor to supplier – breaches of legal obligations and responsibilities can be addressed satisfactorily through a series of actions. Providing, of course,

- there is consistency of terms in the contracts, thereby maintaining continuity of breach
- continuation of loss is maintained and is not broken by indemnities or other terms of trade
- the chain is not broken by insolvency or the like.

Consistency of terms is by no means easy to achieve even when desired but in any event consistency is often deliberately avoided to achieve commercial advantage. The law reports are full of cases where inconsistency has defeated an action and *E R Dyer Ltd* v. *The Simon Build/Peter Lind Partnership* (1982) is of particular interest to civil engineers. The case concerned the relationship between clause 16 of the FCEC blue form of sub–contract and clause 63 of ICE Conditions (Fourth edition). The court held that the expulsion of the main contractor from the site under clause 63 did not constitute determination within the meaning of the phrase 'if the Main Contract is determined for any reason...' in clause 16 of the sub-contract. Consequently the sub-contractor's rights to recover all losses and expenses including profit from the main contractor were not restricted by the provisions of clause 16.

As to indemnity against loss given by one party to another, the case of *North West Metropolitan Regional Hospital Board* v. *T A Bickerton & Son Ltd* (1970) exposed the dangers of the indemnity practices at that time common in nominated sub-contracting. But although such indemnities have been largely eliminated from the ICE Conditions (Fifth and Sixth editions) they are still to be found in some standard forms such as the IChemE Red Book.

Collateral contracts

A collateral contract relates to a warranty given to or given by a party who is not a party to the principal contract. If its existence can be established it forms a useful device for avoiding the doctrine of privity and bridging a gap in a contractual chain.

Various attempts have been made to argue that nomination creates a collateral contract between the employer and the sub-contractor on the basis that in consideration for being nominated the sub-contractor warrants performance to the employer. As a general proposition the argument fails but there are two construction cases on record in respect of named suppliers, *Shanklin Pier Ltd* v. *Detel Products Ltd* (1951) and *Greater London Council* v. *Ryarsh Brick Co. Ltd* (1986), where the courts have accepted the existence of collateral contracts because of the particular circumstances.

Collateral warranties

Collateral warranties are discussed in Chapter 4 in respect of design responsibility. Such warranties are formally executed undertakings which create direct contractual relationships which would not otherwise exist between building owners, leaseholders, tenants, employers, contractors, sub-contractors, suppliers, etc according to how widely they are drafted.

With the opportunities for recovering economic loss under the law of tort severely restricted in the wake of the *D & F Estates Ltd* v. *Church Commissioners for England* (1988) and the *DOE* v. *Thomas Bates* (1990) cases the pressures for collateral warranties are correspondingly increased. So in the building industry it is now routine for sub-contractors and suppliers to be asked to give collateral warranties on design and build projects.

There is nothing in the D and C Conditions directly on this matter but it can be expected that some employer's requirements will seek to impose warranties on sub–contractors and suppliers. And seek is probably the operative word because it has to be said that in the building industry whilst many are asked for not all are given.

10.3 Assignment and sub-contracting distinguished

Starting from the basic position that assignment concerns the transfer of legal rights and sub-contracting concerns delegation or vicarious performance of obligations it can be seen that assignment is effectively an exception to the doctrine of privity of contract whereas sub–contracting gives no relief from the doctrine. Sub-contracting does not excuse the contractor from his obligations to the employer and it does not create legal ties between the employer and sub-contractor.

This straightforward approach which appears to be that adopted by the draftsmen of most standard forms of contract can run into difficulty when subjected to legal scrutiny. In the *Linden Gardens Trust* case mentioned earlier in this chapter the Court of Appeal had to consider the meaning of the following clauses of JCT 63:

- 17(1) 'The Employer shall not without the written consent of the Contractor assign this Contract'
- 17(2) 'The Contractor shall not without the written consent of the Employer assign this Contract, and shall not without the written consent of the Architect (which consent shall not be unreasonably withheld to the prejudice of the Contractor) sub-let any portion of the Works.'

The majority view of the Court of Appeal was that clause 17(1) did prevent the employer from assigning the benefit of the contract without the contractor's consent but it did not prevent the employer from assigning benefits arising under the contract such as accrued causes of action for damages

The dissenting view of Lord Justice Staughton was that the words 'assign this Contract' prohibited only vicarious performance by the employer without consent and did not prohibit assignment of the benefit of the contract or part of it.

The House of Lords upheld the majority view that the contract did prohibit assignment by the employer of the benefit of the contract.

Benefits and burdens

Legal analysis of assignment and sub-contracting is frequently explained in terms of benefits and burdens. As a general rule of English law, benefits can be assigned but not burdens.

This is how Lord Justice Staughton in the *Linden Gardens* case distinguished between the three basic concepts of novation, assignment and sub-contracting:

'(a) Novation

This is the process by which a contract between A and B is transformed into a contract between A and C. It can only be achieved by agreement between all three of them, A, B and C. Unless there is such an agreement, and therefore a novation, neither A nor B can rid himself of any obligation which he owes to the other under the contract. This is commonly expressed in the proposition that the burden of a contract cannot be assigned, unilaterally. If A is entitled to look to B for payment under the contract, he cannot be compelled to look to C instead, unless there is a novation. Otherwise B remains liable, even if he has assigned his rights under the contract to C.

Similarly, the nature and content of the contractual obligations cannot be altered unilaterally. If a tailor (A) has contracted to make a suit for B, he cannot by an assignment be placed under an obligation to make a suit for C, whose dimensions maybe quite different. It may be that C by an assignment would become entitled to enforce the contract – although specific performance seems somewhat implausible – or to claim damages for its breach. But it would still be a contract to make a suit that fitted B, and B would still be liable to A for the price.

(b) Assignment

This consists in the transfer from B to C of the benefit of one or more obligations that A owes to B. These may be obligations to pay money, or to perform other contractual promises, or to pay damages for a breach of contract, subject of course to the common law prohibition on the assignment of a bare cause of action.

But the nature and content of the obligation, as I have said, may not be changed by an assignment. It is this concept which lies, in my view, behind the doctrine that personal contracts are not assignable.

Thus if A agrees to serve B as chauffeur, gardener or valet, his obligation cannot by an assignment make him liable to serve C, who may have different tastes in cars, or plants, or the care of his clothes.

There is no reason in principle why a party who has earned his fee for performing a personal contract should not assign the right to receive it; nor, so far as I can see, why B for whom the tailor has completed a suit should not assign to C the right to receive it

(c) Sub-contracting
I turn now to the topic of sub-contracting, or what has been called in this and other cases vicarious performance. In many types of contract it is immaterial whether a party performs his obligation personally, or by somebody else. Thus a contract to sell soya beans, by shipping them from a United States port and tendering the bill of lading to the buyer, can be and frequently is performed by the seller tendering a bill of lading for soya beans that somebody else has shipped. On the other hand a contract to sing the part of Hans Sachs at Covent Garden Opera House will not be fulfilled by procuring someone else to do so. That is not because the burden of a contract may not be assigned unilaterally; in each case the original contractor would still be liable if the obligation were not performed or were performed badly. It is because some contractual obligations are personal; they must be performed by the party who has contracted to perform them and nobody else.'

10.4 Restrictions on assignment

The D and C Conditions in common with most other standard forms place express restrictions on assignment without consent. The legal effectiveness of such restrictions, however, is not without question. As long ago as 1908 in the case of *Tom Shaw & Co.* v. *Moss Empires Ltd* Mr Justice Darling said this of a prohibition on assignment:

> 'it could no more operate to invalidate the assignment than it could interfere with the laws of gravitation.'

And Lord Justice Staughton again in the *Linden Gardens* case said:

> 'If it were free from authority, I should be inclined to think that a prohibition on assignment in a contract was ineffective. Seeing that assignment cannot increase the obligation or alter it in any way, but only change the person who is to benefit from it, there are no very powerful reasons for allowing it to be prohibited.'

The authority referred to above was principally the decision in the civil engineering case of *Helstan Securities Ltd* v. *Hertfordshire County Council* (1978). A contractor under the ICE Fourth edition had assigned to Helstan amounts due for work undertaken for the employer, Hertfordshire. It was held that Helstan could not recover since the contract provided that the contractor should not 'assign the contract or any part thereof or any benefit or interest therein or thereunder without the written consent of the employer'.

In the *Linden Gardens* case the Court of Appeal resolved the dilemma by holding that a prohibition simply on assigning the contract did not prohibit the assignment of benefits under the contract. The House of Lords took a more robust view and held that:

> 'an assignment of contractual rights in breach of a prohibition against such assignment is ineffective to vest the contractual rights in the assignee.'

These are some of the points made by Lord Browne-Wilkinson in addressing the issue of whether a prohibition on assignment is void as being contrary to public policy.

> 'It was submitted that it is normally unlawful as being contrary to public policy to seek to render property inalienable. Since contractual rights are a species of property, it is said that a prohibition against assigning such rights is void as being illegal.'

> 'In the face of authority, the House is being invited to change the law by holding that such a prohibition is void as contrary to public policy. For myself I can see no good reason for so doing. Nothing was urged in argument as showing that such a prohibition was contrary to the public interest beyond the fact that such prohibition renders the chose in action inalienable.'

> 'A party to a building contract, as I have sought to explain, can have a genuine commercial interest in seeking to ensure that he is in contractual relations only with a person whom he has selected as the other party to the contract. In the circumstances, I can see no policy reason why a contractual prohibition on assignment of contractual rights should be held contrary to public policy.'

> 'The existing authorities establish that an attempted assignment of contractual rights in breach of a contractual prohibition is ineffective to transfer such contractual rights. I regard the law as being satisfactorily settled in that sense. If the law were otherwise, it would defeat the legitimate commercial reason for inserting the contractual prohibition viz, to ensure that the original parties to the contract are not brought into direct contractual relations with third parties.'

Assignment of monies due

Many standard forms of contract seek to avoid dispute on whether the assignment of monies due is caught by express restrictions on assignment by adding into the assignment clause a phrase such as 'save that the Contractor may without such consent assign either absolutely or by way of change any money which is or may become due to him under the contract'.

The ICE standard forms including the D and C Conditions do not follow this approach although it is known that many employers do amend the Conditions to include a similar phrase.

10.5 Restrictions on sub-contracting

The usual question with sub-contracting is not whether the sub-contractor can sue the employer for the value of work he has done – clearly he cannot – but whether the employer can refuse to pay the contractor for work which has been sub-contracted without consent.

The legal position on this was reviewed in the case of *Southway Group Ltd* v. *Wolff and Wolff* (1991). The matter in appeal concerned whether the contractor was entitled to perform his obligations vicariously and whether the employer was obliged to accept vicarious performance as good performance under the contract. It was held that there was an element of personal confidence in the selection of the contractor and that he had no right to sub-contract. Accordingly, the employer was not obliged to accept performance by others.

This is how Lord Justice Bingham reviewed the law:

> 'It is in general permissible for A, who has entered into a contract with B, to assign the benefit of that contract to C. This does not require the consent of B, since in the ordinary way it does not matter to B whether the benefit of the contract is enjoyed by A or by a third party of A's choice such as C. But it is elementary law that A cannot without the consent of B assign the burden of the contract to C, because B has contracted for performance by A and he cannot be required against his will to accept performance by C or anyone other than A. If A wishes to assign the burden of the contract to C he must obtain the consent of B, upon which the contract is novated by the substitution of C for A as a contracting party.
>
> It does not, however, follow that in the absence of a novation A

must personally perform all the obligations he has assumed under his contract with B. In some classes of contract, as where B commissions A to write a book or paint a picture or teach him to play the violin, it would usually be clear that personal performance by A was required. In other cases, as where A undertakes to repair B's shoes or mend B's watch or drive B to the airport, it may be open to A to perform the contract vicariously by employing the services of C. In this situation the contractual nexus remains unaltered, since A remains liable to B for performance of the contract and no contractual relationship arises between B and C.

Whether a given contract requires personal performance by A, or whether (and if so to what extent) A may perform his contractual obligations vicariously, is in my opinion a question of contractual construction. That does not mean that the court is confined to semantic analysis of the written record of the parties' contract, if there is one. Such is not the modern approach to construction of a commercial contract. It means that the court must do its best, by reference to all admissible materials, to make an objective judgment of what A and B intended in this regard.'

Application to the construction industry

The learned editors of Building Law Reports commenting on the *Southway* case say this in 57 BLR at page 35:

> 'It might be thought that these principles [referring to the above review] would not have much application to the modern construction industry. Yet choices are frequently made on the basis of reputation, e.g. of an architect to design a building or an extension to an existing building or of a contractor or a subcontractor on the basis of specialist skill or generally dependable personal performance. It may even be present when competitive tenders are obtained since those selected for tender may have been selected because of the confidence placed in them. In these instances the degree to which sub-contracts may be permissible in law, in the absence of contrary contract provisions, may be severely limited.'

Contractual restrictions on sub-contracting

It can be seen from the above that in the absence of express restrictions on sub-contracting the test for permissible sub-contracting is whether, from the terms of the contract and its circumstances, it can be inferred that it is a matter of indifference whether performance is sub-contracted or not.

Where there are express restrictions, as there are in most standard forms, to the effect that consent must be obtained before sub-contracting, the question then is what is the employer's remedy for the breach. Much depends upon the terms of the contract. Some contracts like the ICE Fifth edition make unauthorised sub-contracting a ground for determining the contractor's employment; others such as the IChemE Green Book state expressly that the employer is not obliged to reimburse the contractor for work which is sub-contracted without approval.

Where there are no express sanctions in the contract for sub-contracting without the approval or consent which is stated to be required the employer's remedy may be no better than damages for breach and these would, of course, have to be proved. It would seem unlikely that the employer could refuse to pay the contractor for work properly executed since by inclusion of a provision for sub-contracting, albeit with consent, the basic legal argument for not paying is destroyed.

10.6 Assignment in the Design and Construct Conditions

Clause 3 of the D and C Conditions is identical in wording to clause 3 of the ICE Sixth edition. Neither party is to assign the contract or any part or any benefit or interest without the prior written consent of the other party. Such consent is not to be unreasonably withheld.

Assignment without consent

Express remedies for assignment without consent are to be found in clause 64(4)(b) – default of the employer – and clause 65(1)(a) – default of the contractor.

Clause 64(4)(b) entitles the contractor to terminate his own employment under the contract for the employer's breach. Clause 65(1)(a) entitles the employer to expel the contractor from the site by notice of termination for the contractor's breach. Both clauses which are similarly worded are surprisingly broad in their coverage – 'assigns or attempts to assign the Contract or any part thereof or any benefit or interest thereunder'. Firstly, there is the point that 'attempts' to assign presumably means 'purports' to assign. There is nothing in clause 3 which makes it a breach of contract to take steps towards assignment in an attempt to assign – it is only the assignment itself which requires consent. Secondly, unlike the ICE Sixth edition where the determination provision relates only to assigning 'the Contract' here the provisions relate to 'the Contract – or any part – or any benefit – or interest thereunder'.

The legal implications of these points may take some time to digest. It is one thing to exclude the common law right to assign a benefit under a contract thereby making such assignment inoperative. It is quite another to elevate the consequence of breach to determination.

'Which consent shall not unreasonably be withheld'

Clause 3 closes with the phrase 'which consent shall not unreasonably be withheld'. The phrase first appeared in clause 3 of the ICE Sixth edition as an addition to clause 3 of the ICE Fifth edition.

There is a view that the phrase is not only unnecessary but potentially troublesome and it is regularly being deleted from ICE Sixth edition contracts. The problem is that the phrase can be taken to imply that under clause 3 the parties have a right to assign and that the unreasonable withholding of consent is itself the breach.

This is clearly not compatible with the termination provisions in clauses 64 and 65 but since the deletion avoids the difficulty it is probably best undertaken.

10.7 Sub-contracting in the Design and Construct Conditions

The general scheme for sub-contracting in the D and C Conditions is that the contractor should be free to select his own sub-contractors without reference to the employer's representative or to the employer. This follows the philosophy adopted in the ICE Sixth edition that under modern commercial arrangements sub-contracting is sufficiently commonplace as to be regarded as normal.

However, not all engineers and employers are happy with this and one of the common amendments being made to the ICE Sixth edition is to replace clause 4 with clause 4 from the ICE Fifth edition which does require the contractor to obtain consent to sub-contracting.

If a similar approach is taken towards the D and C Conditions those making the change will have to give some thought as to how far 'construction' and 'design' need separate treatment.

The whole of the works

Clause 4(1) states that the contractor shall not sub-contract the whole of the construction of the works without the prior written consent of the employer. There is no reference, as in clause 3, to this consent not being unreasonably withheld so it is absolutely at the discretion of the employer how he deals with any request. Note that it is the consent of the employer which must be given – not that of the employer's representative. This raises an interesting point on how far the employer's representative can act as agent of the employer in giving any such consent.

A further point to note is that it is the 'construction' of the works to which clause 4(1) applies. This is a word not used and not necessary in the ICE Sixth edition but in the D and C Conditions it is clearly necessary to avoid the possibility that the contractor might claim that providing design is not sub-contracted he is free to sub-contract the whole of the construction.

There is still the unresolved question of whether sub-contracting the whole of the construction means a single sub-contract package or a series of packages amounting to the whole. It probably means both but there seems to be nothing preventing the contractor from sub-contracting just short of the whole.

The contractor's designer

In line with most standard design and construct forms of contract the employer retains some control over who is to design the works under D and C Conditions. But the control amounts only to consent to changes to designers named in the tender and beyond that the Conditions are silent. There is no requirement for the contractor to give notice of or to seek consent for designers who are not so named.

There is a strong probability that many users of the Conditions will want to see a strengthening provision included.

Clause 4(2)(a) states that the contractor must obtain the employer's consent before making any change in the designer named in the appendix to the form of tender. The clause states that such consent shall not be unreasonably withheld.

Section 5 of Part 2 of the appendix requires the contractor to name his appointed designer and to list those elements for which he has not yet appointed a designer. Since this information is to be given by the tenderer at a time when he will not know whether or not his tender is successful it may be that at this stage any appointment is no more than provisional.

Nevertheless, whether appointed or simply provisionally appointed, it is the designer who is named in the appendix to which clause 4(2)(a) applies.

It can be assumed for practical reasons that where the contractor intends to use in–house design facilities he is expected to name his own firm in the appendix and not the individual designers.

The clause does not deal directly with the situations, commonplace in building but perhaps less likely to be so in civil engineering, where the employer's designer is novated to the contractor or where the designer is named in the employer's requirements. If a designer is so named it is doubted if the contractor has any right to make a change – unless, of course, the employer consents.

Any part of the works

Clause 4(2)(b) permits the contractor to sub-contract 'any part of the construction of the Works'. The only stipulation is that the extent of the work sub-contracted and the name and address of any sub-contractor must be notified to the employer's representative in writing prior to entry to the site.

The opening words of clause 4(2)(b) 'Except where otherwise provided' were thought to refer in the ICE Fifth and Sixth editions to nominated sub-contractors. Presumably here they refer to sub-contractors named in the employer's requirements.

Failure to give notice

There is no express sanction in the Conditions for failing to give notice of sub-contracting. It is most unlikely that the employer could refuse to pay for any work executed to the required standard or that the employer could prove any damage from the breach of failing to give notice. The provision in clause 65(1)(j) – 'persistently or fundamentally in breach of his obligations' – might apply in extreme cases of wilful refusal to give notice of, or to name sub-contractors. This would put the contractor in jeopardy of having his employment terminated.

Labour-only sub-contractors

Clause 4(3) confirms that labour-only sub-contractors need not be notified.

The question is sometimes asked – does this apply to sub-contractors who are essentially labour-only but who work with their own plant? The answer is probably that it does not. There may be a grey area between what are small tools and what is plant but labour-only should be taken to mean what it says and not stretched to include any sub–contractor on the dubious basis that he is labour-only providing that he does not supply his own materials.

The suggestion has been made, somewhat ambitiously it is thought, that designers are labour-only sub-contractors and consequently their employment need not be notified. The point is academic. The Conditions do not require that notice be given for designers whatever their status unless they are changed from named designers. The requirements for notice apply only to 'construction'.

Contractor liable for sub-contractor's work and performance

Clause 4(4) states that the contractor is liable under the contract for the work, acts, defaults and neglects of his sub-contractors, agents, servants or workpeople.

This may not be intended as anything more than a reminder of the ordinary legal position but it may have greater significance. The question is, do the words 'liable under the Contract' cover acts of tort by a sub-contractor as well as contractual defaults?

The answer is probably yes – but only in relation to dealings between the employer and the contractor. That is to say, the clause would not make the contractor liable to third parties for sub-contractor's torts. In law, a sub-contractor is generally regarded as an independent contractor responsible for his own negligence.

There appears to be a small printing error in clause 4(4) in that 'neglects or any sub–contractor' should probably read 'neglects of any sub-contractor'.

Removal of sub-contractors

Clause 4(5) empowers the employer's representative to order the removal of any sub–contractor for misconduct, incompetence, negligence or breaches of health and safety rules.

This is an additional power to that given in clause 16(2) for the removal of the contractor's employees for similar conduct. Clause 16(2) clearly relates to individuals whereas clause 4(5) covers firms.

Under clause 4(5) the employer's representative must first give a warning in writing. He can then serve notice of removal – presumably again in writing but not stated. No time limits are set for the warning or removal but it is suggested that they would have to be whatever was reasonable in all the circumstances.

Failure by the contractor to comply with an order for removal would be a serious breach entitling the employer's representative to consider action under either clause 40 (suspension) or clause 65 (termination of the contractor's employment).

From its context clause 4(5) is clearly intended to apply principally to sub-contractors engaged in the construction of the works. Whether or not it could be used for removal of a design firm is a point for debate. The clause does say from the 'Works', which includes design, and not from the 'Site'. And elsewhere in clause 4 a distinction is made between 'construction' and 'design' where it is thought necessary to make the distinction.

10.8 Terms of sub-contracts

The D and C Conditions do not attempt to regulate the terms on which sub-contracts should be let except that clause 54(7) requires provisions equivalent to the main contract for the payment of goods and materials not delivered to the site.

Some forms, GC/Works/1(edition 3) for example, set out in detail the provisions which must be contained in sub-contracts; others, The Department of Transport design and construct form, for example, require sub-contracting to be on terms which are approved.

There are certainly some provisions in the D and C Conditions, not least those on quality assurance and conciliation, which need to be reflected in sub-contracts to safeguard the interests of the parties.

It is understood that the Federation of Civil Engineering Contractors is producing an amended Blue Form for use with the D and C Conditions.

Chapter 11

Care of the works and insurances

11.1 *Introduction*

Clauses 20 to 25 of the Conditions outline the responsibilities of the parties for accidents and damage and detail the minimum insurance cover which the contractor is obliged to maintain. Thus:

clause 20 – care of the works
clause 21 – insurance of the works
clause 22 – damage to persons and property
clause 23 – third party insurance
clause 24 – accidents to workpeople
clause 25 – evidence and terms of insurance.

Effect of contractor's design

All the above clauses follow closely the wording of the ICE Sixth edition and the principal effect of the contractor taking responsibility for design is likely to be that in practice the excepted risks and exceptions in clauses 20(2)(b) and 22(2)(d) will give the contractor less opportunity to avoid liability than in traditional contracting.

So whereas under the ICE Sixth edition damage to the works due to a design fault or third party damage which is the unavoidable result of the construction of the works is not generally the contractor's liability the position in the D and C Conditions is that such exceptions only apply to design faults for which the contractor is not responsible under the contract. And since the contract seeks very positively to place all design responsibility on the contractor it follows that these particular exceptions have been severely emasculated.

Complexities of interpretation

Although the draftsmen of the D and C Conditions have tried to stay with wording which is well used and tested through various editions of traditional ICE contracts this does not make the complexity of interpretation and application any easier. The fact remains that all insurance clauses enter an area of law which is very much the province of the specialist.

The best advice to contractors and employers alike is follow the instructions of your policies; give prompt notice to your insurers; admit nothing; make no promises; and never negotiate without the consent of your insurers.

The Conditions perhaps suggest otherwise and certainly in the matter of apportionment are an invitation to errors. Apportionment of risk and responsibility is a complex business and not one to be approached on a do-it-yourself basis.

11.2 Professional indemnity insurance

Absence of express provisions

Many comments of surprise have been expressed at various seminars on the D and C Conditions that they contain no provisions for professional indemnity insurance. The thought appears to be that because a designer engaged by the employer is usually required to have such insurance then design and construct contracts should seek to ensure that the contractor or his designer has similar cover. There are various reasons, however, why it is appropriate that the Conditions should be silent on the point.

Firstly, as a general principle the Conditions, and indeed most other construction contracts, deal only with arrangements between the employer and contractor and do not seek to regulate how the contractor should conduct his business with his sub-contractors or suppliers. If a contractor takes on a sub-contractor or a designer who lacks insurance cover it may be unwise but it will rarely be a breach of contract. Then there is the matter of privity of contract and the question of what benefit such subsidiary insurance would be to the employer.

On the broader front there is the point that when an employer engages a designer directly the only effective financial cover for negligence in the design may be professional indemnity insurance. The designer's own financial resources may be inadequate or inaccessible. In contrast, in design and construct contracting the financial status of the contractor will be examined in the selection of the tender list and his ability to stand claims will be taken into account.

Finally there is the point that professional indemnity insurance is usually on a claims made basis so that the cover is only effective in respect of the year in which the claim is made. Thus once a policy lapses there is no cover for past work. Consequently a contractual requirement for such insurance would need to be drafted to ensure that cover is maintained for the legal limitation period rather than merely the construction period as for other insurances. That raises the question – how realistic is it to monitor the maintenance of such insurance over a protracted period of years.

Collateral warranties

If the employer does wish to have cover for design liability over and above that available directly from the contractor then a collateral warranty from the designer may be the solution. Such warranties are discussed in Chapter 4.

Professional indemnity for consultants

Notwithstanding the absence of express contractual requirements in the D and C Conditions for professional indemnity insurance most contractors will insist for their own protection that consultants providing design services have such insurance. This is little different from the policy applied generally in respect of professionals and sub-contractors.

However since the legal responsibility of a professional designer is limited at common law to the exercise of reasonable skill and care and his professional indemnity cover is similarly limited a contractor facing a claim on a fitness for purpose basis will have no redress against the professional designer and no access to his insurance.

It is not unknown for the contractual terms of engagement of a consultant to impose a wider responsibility than reasonable skill and care but there are implications in this – not least the possibility of invalidating the consultant's professional indemnity insurance policy.

Contractor's in-house design insurance

Contractors who undertake in-house design can insure against the negligence of their own designers. The cover is usually defined as being in respect of a negligent act, error or omission of the contractor in performance of his professional activities. But see the *Wimpey* case below.

The need for such insurance arises because a contractor's all risk policy usually excludes design entirely or limits the indemnity to damage caused by negligent design to third party property or construction works other than those designed.

An ordinary professional indemnity policy does not cover the contractor against the problem of discovery of a design fault before completion. At that stage there is no claim against the contractor as there would be against an independent designer. To overcome this contractors usually seek a policy extension giving first party cover. In effect this amounts to giving the construction department of the contractor's organisation a notional claim against the design department.

The complexities of in-house design claims were revealed in the case of *Wimpey Construction Ltd* v. *Poole* (1984) which concerned a contract under the ICE Fourth edition modified for contractor's design. During the construction of a new quay wall at Vospers Southampton shipyard movement of the quay wall occurred and extensive remedial works were necessary. The cause was found to be errors by Wimpey's in-house designers in their assumptions on soil mechanics. Wimpey sought to recover the cost of the remedial works from their insurers and set out to prove their own negligence even to the point of advancing the argument that a company of their standing should be judged by a more stringent and exacting task than an ordinary practitioner.

Wimpey failed to establish negligence although they did win the argument that the words 'negligent act, error or omission' should be construed to include any error or omission without negligence.

Quite apart from the findings of the court, the *Wimpey* case will long be remembered as a cautionary tale on how far the evidence of a particularly brilliant expert witness is relevant in establishing the standards to be expected of an ordinary practitioner. This is part of what Mr Justice Webster had to say about a consultant of world renown who gave expert evidence on behalf of Wimpey:

> 'He is without doubt an outstandingly brilliant exponent of the complexities of soil mechanics and his work in that field has received international acclaim and recognition. He applies,

however, both to himself and to others, the highest possible standards; and it can be fairly said that he is generous with his criticism. Few people connected with the case escaped it. For these reasons, and because his experience has given him little contact with the ordinary day to day problems of designing structures in soil, I am able to place little if any reliance upon his evidence as to the standards to be expected of an ordinarily competent designer'.

11.3 *Insurance terminology*

Insurance clauses in contracts often use phrases which are not particularly clear in themselves but which have particular meanings to insurers. These are a few such phrases.

Subrogation

This is the legal right of an insurer who has paid out on a policy to bring actions in the name of the insured against third parties responsible for the loss.

All risks

An all risks policy does not actually cover all risks since invariably there will be exceptions. However the effect of an all risks policy is to place on the insurer the burden of proving that the loss was caused by a risk specifically excluded from cover. In contrast, under a policy for a specified risk it is the insured who must prove that his loss was caused by the specified risk.

Joint names

Insurance in joint names provides both parties with rights of claim under the policy and it prevents the insurer exercising his rights of subrogation one against the other.

Cross liability

The effect of a cross liability provision in a policy is that either party can act individually in respect of a claim notwithstanding the policy being in joint names.

11.4 *Clause 20 – care of the works*

The contractor's responsibility for care of the works until they are completed is founded in his obligation to design, construct and complete the works. That is an absolute obligation save in so far as it is legally or physically impossible or the Conditions otherwise provide release – for example, employer's default.

Purpose of clause 20

Clause 20 restates this responsibility and then goes further in:

- fixing the time limits during which the contractor carries the risk of care of the works
- stating those risks which are excepted from the contractor's responsibility
- providing how the costs of rectification of loss or damage are to be borne.

Time limits

Clause 20(1)(a) states the broad time limits of the contractor's responsibility as being from the commencement date until the date of issue of a certificate of completion for the whole of the works. Thereafter responsibility passes to the employer.

Note that responsibility expires on the date of issue of the certificate and not on the date for completion on the certificate. This avoids the employer being landed with retrospective responsibility. There is, however, no 14 day period of grace as found in the ICE Fifth edition giving the employer time to arrange his own insurances. On the other hand contractors might regard it as potentially unfair that they remain responsible for care of the works during the 21 day period after completion which is available to the employer's representative under clause 48 for issuing his certificate.

Clause 20(1)(b) deals with the transfer of responsibility when certificates of completion have been issued for parts or sections.

The broad rule again is that responsibility passes to the employer on the date of issue of any certificate. There is, however, a proviso that the contractor remains responsible for damage to such completed work caused by his activities on the site.

Employers need to consider the implications of clause 20(1)(b) when occupying or using parts of the works prematurely since under clause 48(4) the contractor can request a certificate of substantial completion for any such 'substantial' part.

Clause 20(1)(c) adds a further general proviso in respect of outstanding work which the contractor has undertaken to finish during the defects correction period. Responsibility for this work remains with the contractor until it is completed.

Although clause 20(1)(c) applies only to outstanding work itself, responsibility for any damage caused whilst carrying out outstanding work is placed on the contractor under clause 20(3)(a).

Excepted risks

Clause 20(2) defines the risks for which the contractor is not responsible, and which, therefore, fall on the employer. These are:

- use or occupation by the employer or other contractors of any part of the permanent works
- any fault, default, defect etc in design for which the contractor is not responsible
- riot, war etc
- civil war, revolution etc
- radiations, radioactivity etc
- pressure waves from supersonic speeds.

The first two items above imply some involvement of the employer and to the extent that the employer is the cause of any damage he must take legal responsibility. The remaining items fall into a different category. They imply no fault on the part of the employer. They are, in fact, standard exclusions from insurance policies and purely as a matter of balance of risk in the contract these risks are placed on the employer.

Note that under clause 20(2)(a) relating to use or occupation the excepted risk only applies to loss or damage to the 'Permanent Works'. This suggests that if the employer or other contractors use the temporary works they either do so with the contractor's approval and responsibility remains with the contractor or the contractor has to look for his remedy for any loss or damage elsewhere in the contract – perhaps under clause 31 (facilities for other contractors).

Design for which the contractor is not responsible under the contract

The phrase 'design for which the Contractor is not responsible under the contract' appears in the excepted risks of clause 20(2)(b) and the exceptions of clause 22(2)(d). Considering that the Conditions are drafted with the aim of putting full design responsibility on the contractor it is difficult to find many practical applications for the phrase.

One possibility is that if the contractor and the employer's representative cannot agree under clause 8(2)(b) on proposed modifications to design included in the employer's requirements then responsibility for such design may not pass to the contractor.

Rectification of loss or damage

Clause 20(3) sets out in three sub-clauses the consequences which follow from the allocation of risk for care of the works as described in clauses 20(1) and 20(2).

Under clause 20(3)(a) the contractor has to rectify at his own cost any loss or damage for which he is responsible 'so that the Permanent Works conform in every respect with the provisions of the Contract'.

Under clause 20(3)(b) the contractor can be required to rectify any loss or damage arising from the excepted risks at the expense of the employer but only 'if and to the extent required by the Employer's Representative'.

Clause 20(3)(c) deals with apportionment of the cost of rectification when the loss or damage arises from both an excepted risk and a contractor's risk. The provision is in full accord with common sense but it is not free from legal difficulty or the complexities of insurance rules. Generally where a loss has two causes, one defined as a risk and the other as an exception, an insurer is not liable to pay. The apportionment provision may be intended to overcome this. It is also worth noting that by definition in clause 20(2) an excepted risk already assumes apportionment of cause.

Clause 20(3)(c) does not expressly require the contractor to rectify loss or damage caused by both contractor's risk and excepted risk but this can probably be implied.

11.5 Clause 21 – insurance of the works

Clause 21(1) requires the contractor to insure against loss or damage to the works in the joint names of the contractor and the employer. The purpose of this is to ensure that the contractor has funds to complete the works in the event of damage and alternatively to protect the employer's investment in the works.

The precise phrase in clause 21(1) is not the works but 'the Permanent and Temporary Works'. This highlights the distinction between the physical works on site and 'the Works' which by definition includes design.

Joint names

The effect of putting insurance in joint names is to permit the employer to claim directly on the policy but it also has the advantage of preventing the insurer from exercising rights of subrogation between the parties.

Replacement cost plus 10%

The insurance under clause 21(1) is to cover full replacement cost plus an additional 10% for incidentals, fees, demolition etc.

Extent of insurance cover

Clause 21(2) regulates the extent and duration of insurance cover. It must:

- cover all loss or damage from whatsoever cause arising other than the excepted risks
- run from the commencement date to the date of issue of the relevant certificate of substantial completion
- cover loss or damage arising during the defects correction period from a cause arising prior to the issue of any certificate of substantial completion

- cover loss or damage caused by the contractor in carrying out any work during the defects correction period.

Defective workmanship

Clause 21(2)(c) confirms that the contractor is not obliged to insure against defective work. This is usual if for no better reason than the fact that such insurance is not readily available.

The clause does end with the phrase 'unless the Contract otherwise requires'. Presumably such a requirement would be stated in the employer's requirements or an amendment to the Conditions. In the ICE Sixth edition the corresponding phrase is 'unless the Bill of Quantities shall provide a special item for this insurance'.

Unrecovered losses

Clause 21(2)(d) is a provision which emphasises the point in clause 21(1) that insurance does not limit the obligations of the parties. It states that amounts not insured or recovered from insurers, whether by excesses or otherwise, shall be borne by the parties in accordance with their respective responsibilities.

This clause needs to be read in conjunction with clause 25 (evidence and terms of insurance) which, amongst other things, allows for controls on excesses.

11.6 *Clause 22 – damage to persons and property*

Clause 22 relates to third party claims. It requires the contractor to indemnify the employer against losses and claims arising out of the design, construction or completion of the works in respect of:

- death or injury to any person, or
- loss or damage to any property other than the works.

The indemnity is to cover all claims, proceedings etc. and all charges and expenses whatsoever. It extends, therefore, to legal costs as well as to sums claimed.

However the case of *Richardson* v. *Buckinghamshire County Council* (1971) under the ICE Fourth edition is worth noting. A motor cyclist sued the County Council for injuries alleged to have been sustained when he fell off his machine at roadworks. The County Council successfully defended the claim but had to meet their own costs because the motor cyclist was legally aided. They tried to recover their costs from the contractor under clause 22 but it was held by the Court of Appeal they had not arisen 'out of or in consequence of the execution of the works'.

Exceptions

As with clause 20 there are exceptions to the contractor's indemnity obligations. Clause 22(2) lists the following items as the responsibility of the employer:

- crop damage
- use or occupation of land
- right of the employer to construct on land
- damage which is the unavoidable result of the construction of the works in accordance with the employer's requirements
- claims arising from negligence or breach by the employer or his agents.

Unavoidable damage

In traditional civil engineering contracts the exception in clause 22(2)(d) for damage which is the unavoidable result of the construction of the works is by far the most commonly invoked and the most contentious exception. Usually it is claimed for damage to property through excavations, dewatering or vibrations.

It is very much in the contractor's interest to make such a claim where circumstances permit because he thereby avoids a claim on his insurance and any delay or extra costs in completing the works may fall on the employer.

In the D and C Conditions clause 22(2)(d) only applies to damage which is the unavoidable result of the construction of the works 'in accordance with the Employer's Requirements including any design for which the Contractor is not responsible under the Contract'. The ICE Sixth edition uses the words 'in accordance with the Contract'.

The intention in the D and C Conditions is probably that the exception should only apply when the contractor is not responsible for design. This would be in keeping with the distribution of responsibility in the Conditions. However it is possible to read the clause as applying to all work in accordance with the employer's requirements which would have the effect of making unavoidable damage resulting from the contractor's design an exception.

An improvement to the wording of clause 22(2)(d) would clearly be welcome.

Indemnity by employer

Clause 22(3) requires the employer to indemnify the contractor against all claims etc made in respect of the exceptions in clause 22(2).

Shared responsibility

Clause 22(4) on shared responsibility contains two separate provisions: one for the contractor; the other for the employer.

In clause 22(4)(a) the contractor's liability to indemnify the employer is reduced in proportion to the extent that the employer or his agents may have contributed to the damage or injury.

In clause 22(4)(b) the employer's liability to indemnify the contractor in respect of the exceptions in clause 22(2)(e) is reduced in proportion to the extent that the contractor or his agents etc may have contributed to the damage or injury.

There is no reduction of employer's liability in respect of the other excepted risks in sub-clauses 22(2)(a), (b), (c) and (d) – presumably on the basis that there can be no element of contractor's fault. This may be of some assistance in the interpretation of clause 22(2)(d) discussed above.

11.7 Clause 23 – third party insurance

Clause 23 provides for insurance against third party risks in much the same way that clause 21 provides for insurance of the works.

Clause 23(1) requires the contractor to insure in joint names against liabilities for death or injury to any person (other than a person in the workforce) and for loss or damage to property (other than the works), arising out of the performance of the contract.

The contractor is not required to insure in respect of the exceptions in clauses 22(2)(a) to (d) but the insurance must cover the exception in clause 22(2)(e). This is because 22(2)(e) relates to the matters specifically addressed in the clause 23(1) insurance.

Cross liability

Clause 23(2) requires the insurance policy to have a cross liability clause. This permits either party to claim as separate insured.

Amount of insurance

Clause 23(3) states, as is usual in civil engineering contracts, that third party insurance is to be for at least the amount stated in the appendix to the form of tender.

11.8 Clause 24 – injury to workpeople

Clause 24 provides that the employer shall not be liable for damages or compensation for accidents or injuries to the contractor's workforce except to the extent that default of the employer is a contributory factor.

The wording is identical to that in the ICE Sixth edition with one exception. The phrase 'save and except and to the extent' replaces 'save and except to the extent'. The additional 'and' appears to have been added to clarify the position on proportional responsibility.

There is no express requirement in clause 24 for insurance cover but employers (in the broad sense) are required by statute to carry insurance against injury to their employees in the course of their employment.

There can be problems with labour-only and other self-employed persons who do not benefit from employer's insurance cover. To guard against this and other claims clause 24 requires the contractor to indemnify the employer against all claims.

11.9 Clause 25 – evidence and terms of insurance

Clause 25 deals with five issues:

- evidence of insurance
- terms of insurance

- excesses on policies
- failure to insure
- compliance with policy conditions.

Evidence and terms

Clause 25(1) requires the contractor to provide satisfactory evidence prior to the commencement date that the insurances are in force. If required the policies must be produced for inspection.

The clause further provides that the terms of the insurance shall be to the approval of the employer and that the contractor shall, on request, produce receipts of payment of premiums. There is no requirement for insurance to be with an insurer approved by the employer.

Excesses

Clause 25(2) provides that any excesses on policies shall be as stated by the contractor in the appendix to the form of tender.

The provision is intended to guard against the contractor taking on insurance with high excesses and thereby leaving the employer exposed to the amount of the excess in the event of the contractor's financial failure. Some employers state in the tender documents the maximum excesses they are prepared to accept.

Failure to insure

Clause 25(3) allows the employer to take out insurance when the contractor fails to produce satisfactory evidence that his insurances are in force. The premiums paid by the employer can be deducted from monies due to the contractor or recovered as a debt.

Contractors would be most unwise to allow this state of affairs to develop since the employer would insure in his own name leaving the contractor exposed to claims from the insurer under his rights of subrogation.

Compliance with policy conditions

Clause 25(4) is a provision requiring both the employer and the contractor to comply with the conditions laid down in the insurance policies and indemnify each other against any claims arising from failure to comply.

The effect of this is that if one party renders the insurance policies void by his actions or omissions, he then stands in the place of the insurer in providing cover.

There is a minor change in wording from the ICE Sixth edition with the phrase 'Should the Contractor or the Employer fail' replacing 'In the event that the Contractor or the Employer fails'. This seems to amount to no more than a drafting correction in recognition of the point that failure to do something is not an event.

Chapter 12

Statutes, facilities and fossils

12.1 *Introduction*

This chapter covers:

clause 26 – giving of notices and conforming with statutes
clause 28 – patent rights and royalties
clause 29 – interference with traffic, noise, disturbance and pollution
clause 30 – damage to highways and transport of materials
clause 31 – facilities for other contractors
clause 32 – fossils etc.

Street works

Clause 27 is not used in the D and C Conditions. In the ICE Sixth edition clause 27 is the lengthy clause dealing with the Public Utilities Street Works Act 1950. That Act has now been replaced by the New Roads and Street Works Act 1991. The D and C Conditions refer to the new Act in clause 26(3)(d) which makes the contractor responsible for the serving of notices under the Act.

Statutory notices

The provisions in clause 26 of the D and C Conditions detailing responsibility for giving notices, conforming with statutes and obtaining planning permission are not significantly different from those in the ICE Sixth edition.

More change might have been expected to take account of the contractor's design responsibilities although, in the very important matter of which party is responsible for obtaining planning permission, some flexibility is introduced. The responsibility falls on the employer unless 'the Contract otherwise provides'.

Patent rights

Clause 28 on patent rights and indemnities against infringements has been modified to place greater liability on the contractor. But, nevertheless, the employer remains liable for any design or specification 'provided other than by or on behalf of the Contractor'.

Noise, disturbance and pollution

Clause 29 is unchanged from clause 29 in the ICE Sixth edition. It is questionable however whether, in a design and construct contract, the employer should have to indemnify the contractor against liability for damages on account of noise, disturbance or other pollution 'which is the unavoidable consequence of carrying out the Permanent and Temporary Works'.

Damage to highways

Clause 30 is also unchanged from the ICE Sixth edition and again the question arises whether the employer should give the same indemnities in a design and construct contract as in a traditional contract.

Facilities for other contractors

Clause 31, requiring the contractor to provide facilities for other contractors remains the same in the D and C Conditions as in the ICE Sixth edition.

In practice the clause will probably have less impact than in the ICE Sixth edition since, in design and construct contracting, the works of other contractors are more likely to be brought within the scope of the main contractor's responsibility.

Fossils

Clause 30 on fossils and antiquities appears to be immune from change in whatever version of ICE Conditions it appears.

However, there is nothing obviously different about design and construct contracting which would have justified re-drafting the clause in the D and C Conditions.

12.2 Clause 26 – notices, fees and statutes

Serving of notices and payment of fees

Clause 26(1) states that the contractor shall give all notices and pay all fees required to be given or paid by:

- any Act of Parliament
- any regulation or bye-law of any local or statutory authority
- the regulations of public bodies and companies whose property or rights are affected by the works.

This is subject to the proviso 'save as provided in sub-clause (3)(d) of this Clause'.

The proviso in sub-clause (3)(d) is written as 'unless the Contract otherwise provides' and it relates to the serving of notices under the New Roads and Street Works Act 1991. An identically worded proviso commences sub-clause (3)(c) relating to planning permission but there is no mention of this in clause 26(1). This may be because obtaining planning permission is not seen as giving a statutory notice or it may simply be a drafting slip in the change from the ICE Sixth edition.

The type of notices which fall under clause 26(1) as a matter of routine include:

- health and safety
- building regulations
- control of work
- demolitions
- use of explosives
- closure of highways
- permits for skips.

In respect of serving such notices there is no significant difference between traditional contracting and design and construct contracting but the implications in the event of delay in consent or refusal can be different.

In traditional contracting the engineer can, for some problems, be drawn into issuing a variation order and/or an extension of time – for example, difficulties with building regulations. But in design and construct contracting the impact of most problems will remain with the contractor without any relief from the provisions of the contract or any reduction in obligations under the contract.

Repayment of fees

Clause 26(2) provides that the employer shall allow or repay the contractor:

- such sums as the employer's representative shall certify to have been properly payable 'and paid', and,
- rates and taxes paid by the contractor in respect of:
  - the site or any part thereof
  - temporary structures situated elsewhere used exclusively for the works
  - any structures used temporarily and exclusively for the works.

This is identical to clause 26(2) of the ICE Sixth edition but there is a good case for saying that it should have been comprehensively revised.

Firstly, with design and construct contracting it is usual for the contractor to include in his price for the payment of fees – particularly where the price is on a lump sum basis. The contractor, by his design, his methods and his competence can influence the amount of such fees and he will seek to keep them to a minimum if they are not refundable.

Secondly, in respect of rates and taxes, the provisions which have been copied from the ICE Sixth edition are, to say the least, uncertain in their application. For example – does the phrase 'any temporary structures situated elsewhere' oblige the employer to pay the rates on the contractor's site compound? Another question is – does the phrase 'any structures used temporarily and exclusively for the purposes of the Works' oblige the employer to pay the rates on any permanent buildings used by the contractor if they are fully given over to the contract? Could this include the rates on design offices in certain circumstances? And, again, since the amounts of rates and taxes payable are to some extent within the control of the contractor why should these costs not be included in the contractor's price?

Conforming with statutes

Clause 26(3) requires the contractor to ascertain and conform in all respects with:

- Acts of Parliament
- regulations and bye-laws of local or statutory authorities
- rules and regulations of public bodies.

The contractor is required to indemnify the employer against all penalties and liability of every kind for breach of any Act, regulation or bye-law.

There are, however, two important exemptions to those duties:

- the contractor is not required to indemnify the employer against any breach which is the unavoidable result of complying with the contract or instructions of the employer's representative – clause 26(3)(a)
- unless the contract provides otherwise the contractor is not responsible for obtaining planning permission – clause 26(3)(c).

Moreover, by clause 26(3)(b), if the employer's requirements preclude conformity with any Act, regulation or bye-law the employer's representative is to order a variation under clause 51.

Contractor not relieved of legal responsibility

The exception in clause 26(3)(a) to the indemnity given to the employer does not, of course, relieve the contractor of his legal responsibility for any breach of statute, regulation or bye-law. All that it does is relieve the contractor from his indemnity to the employer.

In the case of *Eames* v. *North Hertfordshire District Council* (1980) a contractor erected a portal frame on made-up ground. Both the architect and the Council's building inspector allowed the work to proceed. Nonetheless the contractor was held liable for breach of a statutory duty (under the Building Regulations) and liable in negligence. Judge Fay QC referred in his judgment to the comment of Lord Wilberforce in *Anns* v. *London Borough of Merton* (1978):

> 'Since it is the duty of the builder, owner or not, to comply with the bye-laws, I am of the opinion that an action could be brought

against him in effect for breach of statutory duty by any person for whose benefit or protection the bye-law was made.'

The message for contractors is clear enough – comply with the employer's requirements and the instructions of the employer's representative only to the extent that it is lawful to do so.

'Unavoidable result of complying with the Contract'

The wording of the exception in clause 26(3)(a) – 'the unavoidable result of complying with the Contract' – is the same as that in the ICE Sixth edition but it may have wider implications in the D and C Conditions. This is because the 'Contract' by definition in clause 1(1)(g) of the Conditions includes the contractor's submission.

This could leave the employer exposed to liabilities arising from breaches caused by defaults in the contractor's submission. It is doubtful if this is the intention of the Conditions but to remove any doubt it may be necessary to replace 'Contract' with 'Employer's Requirements'.

'Instructions of the Employer's Representative'

The phrase 'instructions of the Employer's Representative' in clause 26(3)(a) follows the phrase 'instructions of the Engineer' in the ICE Sixth edition where, of course, it had more obvious application.

In the D and C Conditions there is not much scope for the employer's representative to give instructions which are not variations – and, therefore, by definition within the 'Contract' as employer's requirements.

Variations to ensure conformity

Clause 26(3)(b) states that if the employer's requirements are at any time found to preclude conformity with any Act, regulation or bye-law, then the employer's representative shall 'issue such instructions including the ordering of a variation under Clause 51 as may be necessary to ensure conformity'.

It is suggested that this clause with its specific wording takes precedence over the generality of clause 8(2)(b) which requires the contractor to accept responsibility for any design in the employer's requirements.

The implications under the D and C Conditions of a change in the employer's requirements to ensure conformity are potentially far reaching. For whereas under traditional contracting the engineer can, so to speak, dig himself out of his own hole, in design and construct contracting the employer's representative is obliged to consult, if not rely on, the contractor in finding a solution.

Again, under clause 26(3)(b) it is not clear how the employer's representative can issue an instruction which is not a variation.

Planning permission

In design and construct contracting the question of which party is responsible for obtaining planning permission is of fundamental importance. There is no point in letting a contract for something which cannot be built.

So at the very least the employer will normally have outline planning permission before he considers going out to tender. He must then decide on a major issue. Does he take the design to a stage where it can obtain detailed planning permission and incorporate that design in his employer's requirements? Or does he leave the tenders, and eventually the chosen contractor, with the flexibility to offer different detailed designs? And if he does this, who takes the risk once the contract is in place if the contractor's design meets with refusal from the planners?

In practice, a great deal depends upon the type of project which is being undertaken. If it is a one-off scheme with high visual impact and environmentally sensitive repercussions it would be folly to proceed to tender stage with anything other than reserved matters outstanding in the planning permission. But for a commonplace scheme such as a factory unit on an industrial estate it is not uncommon for the contractor to take on the risk and the burden of obtaining planning permission.

Planning permission in the Design and Construct Conditions

In the D and C Conditions, clause 26(3)(c) is worded so that the Conditions retain the flexibility to offer either employer's risk and burden or contractor's risk and burden. This is achieved by opening the clause with the phrase 'unless the Contract otherwise provides'.

That phrase allows for the employer's requirements or a special condition of contract to override the general provisions in clause 26(3)(c) which state that:

- the contractor shall not be responsible for obtaining planning permission for the permanent works or for any temporary works designed by others
- the employer is deemed to warrant that all planning permissions have been, or will be, obtained in due time.

One effect of the employer warranting that he will obtain planning permission in due time is that he is liable to the contractor for damages if he fails. The Canadian case of *Ellis-Don Ltd* v. *Parking Authority of Toronto* (1978), much quoted as authority on winter working and the recovery of overheads, concerned the failure of the Authority to obtain building permits in time.

New Roads and Street Works Act 1991

Clause 26(3)(d) of the D and C Conditions replaces clause 27 of the ICE Sixth edition in dealing with notices to be served in respect of streetworks.

The new clause omits the detail given in the ICE Sixth edition of the Public Utilities Street Works Act 1950. It simply refers to the New Roads and Street Works Act 1991 and states that:

- unless the contract otherwise provides
- the contractor shall be responsible for the service of notices under the Act after the award of the contract.

12.3 *Clause 28 – patent rights and royalties*

In all types of construction contract it is usual to require the contractor to indemnify the employer, in some form, against claims in respect of patent rights and royalties for things selected by the contractor for incorporation in the works. But clearly, in design and construct contracts such indemnities are of greater significance than in traditional contracting with engineer's design.

Clause 28(1) of the D and C Conditions requires the contractor to indemnify the employer against all claims and proceedings for infringement of:

- patent rights
- design trademarks
- names
- other protected rights.

The indemnity is to be given in respect of:

- design, construction and completion of the works
- any contractor's equipment used in connection with the works
- any materials, plant or equipment incorporated in the works.

The indemnity, however, is subject to exceptions as provided in clause 28(2).

Clause 28(2) provides that the employer shall indemnify the contractor when the infringement results from:

- compliance with any design or specification provided other than by or on behalf of the contractor
- use of the works, or any part, in association or in combination with any artifact not supplied by or on behalf of the contractor
- use of the works other than for a purpose indicated by, or reasonably to be inferred from, the employer's requirements.

The first of these exceptions – infringement due to compliance with any specification provided other than by the contractor – has possibilities which may go beyond the intentions of the draftsmen. The word 'specification' as used here should, perhaps, relate only to named products but it may be open to a contractor to argue that it relates to any products which comply with a particular specification.

Resolution of this, however, and precisely what is meant by use 'in association with or in combination with any artifact' is best left to specialist lawyers.

Royalties

Clause 28(3) states that the contractor shall pay all tonnage and other royalties, rent and other payments or compensation for getting stone, sand, gravel, etc for the works.

In the ICE Sixth edition the corresponding clause commences with the words 'Except as otherwise stated'. This presumably relates to borrow pits or other sources of materials arranged by the employer. It is not clear why this phrase has been dropped from the D and C Conditions. Clause 28(3) as it remains appears to be no more than a statement of the obvious.

12.4 Clause 29 – interference, noise and pollution

Clause 29 is identical to clause 29 in the ICE Sixth edition. It deals with various aspects of nuisance to third parties as opposed to damage, which is dealt with in clause 22. The first part of the clause deals with traffic, the second part with noise and other pollution.

Traffic

Clause 29(1) requires that all operations for the construction of the works shall be carried out, so far as compliance with the contract permits and so as not to interfere 'unnecessarily or improperly' with:

- the convenience of the public
- access to roads, footpaths, properties
- use of roads, footpaths, properties.

The contractor is required to indemnify the employer against all claims and proceedings 'whatsoever arising'. Note however, that there is a discrepancy between the obligation on the contractor to avoid interference and the indemnity he must give. The obligation is qualified by the phrases:

- 'so far as compliance with the requirements of the Contract permits' and
- 'not to interfere unnecessarily or improperly'.

The indemnity is not so qualified.

This would suggest that even when the contractor is not at fault contractually, in that he is merely complying with the requirements of the contract, he is required to indemnify the employer against claims. Perhaps there should be a counter-indemnity by the employer similar to that given in clause 29(2) as discussed below.

The phrase 'requirements of the Contract' in clause 29(1) is open to the comment, made repeatedly in this book, that the 'Contract' includes the contractor's submission. The intention of the clause seems to be to excuse the contractor from the consequences of complying with specific employer's requirements. But by using the word 'Contract' instead of the words 'Employer's Requirements' the clause may give the contractor far wider latitude if, in his submission, the contractor includes qualifications and the like.

Noise, disturbance and pollution

Clause 29(2) requires that all work shall be carried out without unreasonable noise or disturbance or other pollution.

Clause 29(3) requires the contractor to indemnify the employer against claims and proceedings except to the extent that noise, disturbance or pollution are the 'unavoidable consequence of carrying out the Works'.

Clause 29(4) requires the employer to indemnify the contractor against claims which are so unavoidable.

There is, of course, wide scope for argument when claims arise as to what is 'unavoidable'. And since insurers may be involved in meeting claims, either directly or through indemnities, they will want to be involved in any decisions reached. Indeed, many policies preclude any admission of liability by the parties themselves. The safest course for the parties is to keep their insurers fully informed and to act only on their advice.

12.5 Clause 30 – damage to highways

Clause 30 on damage to highways is another clause virtually unchanged from that in the ICE Sixth edition.

The clause sets out, as between the parties, their respective liabilities under the Highway Act 1980 or, in Scotland, the Roads (Scotland) Act 1984. The Acts allow the highway authority to recover its excess maintenance expenses arising from excessive weight or extraordinary traffic.

Broadly the scheme is that the contractor takes all risk from movement of his equipment and temporary works and the employer takes the risk from materials and manufactured articles.

Clause 30(1) requires the contractor to use every reasonable means to prevent extraordinary traffic and to use suitable routes and vehicles to ensure that any inevitable extraordinary traffic is limited as far as possible and no unnecessary damage is caused to roads and bridges.

Clause 30(2) provides that the contractor must pay the costs of strengthening or improving the roads or bridges to facilitate movement of his equipment or temporary works and must indemnify the employer against all claims.

Clause 30(3) requires the contractor to notify the employer as soon as he becomes aware of any damage caused by the transport of materials or manufactured articles. The employer is to negotiate and settle any claims and to indemnify the contractor. If the claim is due to any failure by the contractor to use reasonable means and limit damage as required by sub-clause (1) then the employer's representative is required to certify the amount due from the contractor to the employer.

As in clause 20, this brings the employer's representative into the field of the loss adjuster, and its application, if not its wisdom, is questionable.

12.6 *Clause 31 – facilities for other contractors*

The primary purpose of clause 31 is to make clear that the contractor is not entitled to sole possession of the site.

The clause obliges the contractor to afford all reasonable facilities, in accordance with the requirements of the employer's representative, to:

- other contractors employed by the employer and their employees
- employees of the employer
- employees of authorities or statutory bodies employed in the execution of work not in the contract.

Omitted from the D and C Conditions is the category found in the ICE Sixth edition – employees of authorities or statutory bodies employed in the execution of any contract which the employer may enter into in connection with or ancillary to the works. The omission is not thought to be of any significance since the categories which remain are sufficiently comprehensive.

Contractor's responsibility

Clause 31 has its origins in traditional ICE contracts where the involvement of 'other contractors' on the site is accepted as normal.

However, in design and construct contracting much of the benefit of single point responsibility can be lost if that situation is not curtailed.

There is something to be said, therefore, for giving the contractor full responsibility for delivering the works and making him responsible for the performance of 'other contractors'. To do so needs contractual provisions which require the contractor to place directly all orders for ancillary works and to include the costs in his price. This is the approach of The Department of Transport's design and construct conditions.

Although design considerations such as co-ordination and integration will encourage this move towards contractor's responsibility there is the added benefit that claims for delay and disruption caused by the 'other contractors' need not necessarily involve the employer. By being in contract with 'other contractors' the contractor has direct remedies against them and the incentive to minimise their impact.

'Reasonable facilities'

The phrase 'reasonable facilities' in clause 31 is open to a variety of interpretations. It probably means no more than allowing access and space in which to work. It may also include the use of scaffolding, messrooms, toilets and the like and possibly the supply of power and water.

As to what is reasonable or unreasonable, the point may be academic. The contractor can either recover his unforeseen cost for that which is reasonable under clause 31(2); or his cost for that which is unreasonable under clause 52 on the grounds that there must have been a change in the employer's requirements constituting a variation.

Delay and extra cost

Clause 31(2) provides that if compliance with affording reasonable facilities involves the contractor in delay or cost beyond that reasonably foreseeable at the date of award of the contract, the contractor is entitled to:

- an extension of time under clause 44
- payment of the amount of such cost as may be reasonable
- profit on any additional work.

12.7 *Clause 32 – fossils etc.*

Clause 32 deals with fossils, coins, articles of value, antiquities or other things of interest found on the site. Perhaps with some logic it has remained unchanged through successive editions of ICE Conditions.

The clause has essentially two functions:

- to deal with ownership as between the contractor and the employer
- to deal with removal and disposal.

As to ownership all things found on the site are deemed to be the absolute property of the employer.

As to discovery and disposal, the contractor is required to take precautions to prevent removal or damage and is to acquaint the employer's representative immediately upon discovery. The contractor is to carry out at the expense of the employer the employer's representative's orders for examination and disposal.

If delay is caused the contractor should have no difficulty obtaining an extension of time under the 'special circumstances' of clause 44.

Chapter 13

Materials, workmanship and suspension of work

13.1 *Introduction*

In the building industry where design and build has a substantial lead time in its use over design and construct in civil engineering, disputes on materials and workmanship are probably the most significant source of contention. This is hardly surprising. The employer, who relies on the contractor's design and selection of materials and who regards the lowest tendered price as the best bargain is more than likely to be disappointed with the finished product. Top quality and the lowest price are not natural bedfellows. Or, in the maxim of the motor trade – you do not get a Rolls Royce for the price of a Mini.

The employer who has previously employed his own designer and specified his requirements in detail has, intentionally or not, set a base standard for quality and a platform for price. Moving to contractor design is a potentially hazardous business. Unless the employer's requirements are clearly drafted neither contractual provisions on quality and workmanship, common law rules nor statutory protection may provide much comfort. Viewed only by such tests as 'appropriate standards', 'codes of practice', 'merchantable quality', and 'fitness for purpose' a Mini may well be an appropriate substitute for a Rolls Royce.

The Design and Construct Conditions

The draftsmen of the D and C Conditions rely heavily on the employer's requirements to set standards for materials and workmanship. The provisions in clauses 36 to 39 follow closely those in the corresponding clauses of the ICE Sixth edition which are, of course, based on the principle of employer's design. If the D and C Conditions are used with design parameters rather than a detailed design brief from the employer, the provisions in the Conditions amount to little more than a fall back position. And this is a position which could prove expensive to the employer in costs for samples and testing.

Suspension of work

The D and C Conditions have retained, as in the ICE Sixth edition, clause 40 on suspension of work under the general heading of materials and workmanship. In fact, there is nothing in the clause which restricts its operation to these matters but since this is the familiar place for the clause in ICE Contracts it is dealt with in this chapter.

As it happens, clause 40 in the D and C Conditions is virtually identical to that in the ICE Sixth edition although one could argue the case for the employer having less right to interfere by suspending work under design and construct contracting than under traditional contracting.

Ownership of materials and equipment

Clause 54 on the ownership and vesting of materials and the contractor's equipment has been brought into this chapter for the sake of completeness. In the ICE Sixth edition both clause 53 and clause 54 apply to these matters but in the D and C Conditions clause 53 is redesignated to additional payments.

13.2 *Materials and workmanship*

The contractor's responsibilities for providing materials and workmanship to a particular standard derive from the express terms of the contract and from terms implied by common law or statute.

Common law

The courts will imply two independent warranties into contracts for work and materials:

- that materials will be of good and merchantable quality
- that materials will be reasonably fit for the purpose for which they are used.

These warranties apply unless the circumstances of the contract can be shown to exclude them.

The general rule was laid down in the case of *Myers* v. *Brent Cross Service Co.* (1934) where Mr Justice du Parcq said:

> 'a person contracting to do work and supply materials warrants that the materials which he uses will be of good quality and reasonably fit for the purpose for which he is using them, unless the circumstances of the contract are such as to exclude any such warranty.'

Two later decisions of the House of Lords show how the general rule may be modified by particular circumstances. In *Young & Marten Ltd* v. *McManus Childs Ltd* (1969) a roofing sub-contractor was held to be responsible for a batch of defective tiles although both the brand and manufacturer of the tiles (Somerset 13) were specified. It was held that the specification of a particular brand excluded the warranty that the tiles would be fit for purpose but the circumstances of the case were not such as to exclude the warranty that the tiles would be of merchantable quality. Since the sub-contractor had fixed tiles with latent defects he was in breach. Lord Pearce had this to say on the circumstances which would exclude a warranty:

> 'If it is known to both parties that the manufacturer gives no warranty to the contractor, that fact is a strong indication that no warranty is being given by the contractor. So, too, of course, if a contractor advises against a particular material. But the circumstances of contracts are so various that it must be a question of fact and degree whether the circumstances of a particular case suffice to exclude a warranty which the general rule implies.
>
> In the present case the employer's choice of Somerset 13 tiles, which were manufactured by only one firm, is not in itself sufficient to exclude the warranty of quality.'

In *Gloucestershire County Council* v. *Richardson* (1969) concrete columns made by a nominated supplier were found to be defective after erection. It was held that the contractor was not responsible since the circumstances of the case indicated an intention to exclude from the main contract any implied terms that the columns would be of good quality or fit for their purpose. This is how Lord Wilberforce explained the matter:

'The situation thus created was one of a special and complex character, differing greatly from that which arose in *Young & Marten Ltd* v. *McManus Childs Ltd*. There the employer nominated a brand article to be supplied by the manufacturer with no limitation on the contractor's freedom to contract with the manufacturer as he thought fit. The contractor could, and it would be the expectation that he would, or at least it would be his responsibility if he did not, deal with the manufacturer on terms attracting the normal conditions or warranties as to quality or fitness.

But here, the design, materials, specification, quality and price were fixed between the employer and the sub-supplier without any reference to the contractor; and so far from being expected to secure conditions or warranties from the sub-supplier, he had imposed upon him special conditions which severely restricted the extent of his remedy. Moreover, as reference to the main contract shows, he had no right to object to the nominated supplier, though, by contrast, the contract does provide a right to object to a nominated sub-contractor if the latter does not agree to indemnify him against his liability under the contract.'

Statute

The common law rules are now partly codified in the Supply of Goods and Services Act 1982. This is titled as:

'An Act to amend the law with respect to the terms to be implied in certain contracts for the transfer of property in goods, in certain contracts for the hire of goods and in certain contracts for the supply of a service; and for connected purposes.'

Supply of Goods and Services Act 1982

Part I of the Act applies to the supply of goods and Part II of the Act applies to the supply of services. A construction contract with its composite provisions for goods and services attracts both parts of the Act.

Under Part I goods must conform with statutory obligations as to quality and fitness as implied by Sections 2 to 5 of the Act unless these obligations are displaced by particular circumstances or excluded by agreement:

- Section 2 covers implied terms about title.
- Section 3 covers implied terms where transfer is by description. In such a case there is an implied condition that the goods will correspond with the description.
- Section 4 covers implied terms about quality or fitness. Where goods are supplied in the course of business there is an implied condition that they are of merchantable quality. Where the intended purpose of the goods is made known, there is a further implied term that they are reasonably fit for that purpose.
- Section 5 covers implied terms where transfer is by sample. Where goods are supplied by reference to a sample there are implied conditions that the bulk will correspond with the sample in quality and that the goods will be free from defects rendering them unmerchantable which would not be apparent from examination of the sample.

In Part II of the Act relating to the supply of services, Sections 13 to 16 deal with implied terms:

- Section 13 – Implied term about care and skill
  In a contract for the supply of a service where the supplier is acting in the course of a business, there is an implied term that the supplier will carry out the service with reasonable care and skill.
- Section 14 – Implied term about time for performance
  Where the time for the service to be carried out is not fixed by the contract, left to be fixed in a manner agreed by the contract or determined by the course of dealing between the parties, there is an implied term that the supplier will carry out the service within a reasonable time.
- Section 15 – Implied term about consideration
  Where the consideration for the service is not determined by the contract, left to be determined in a manner agreed by the contract or determined by the course of dealing between the parties, there is an implied term that the party contracting with the supplier will pay a reasonable charge.
- Section 16 – Exclusion of implied terms
  (1) Obligations under the Act may be negotiated or varied by express agreement, or by the course of dealing between the parties, or by such usage as binds both parties
  (2) An express term does not negate a term implied by the Act unless inconsistent with it
  (3) Nothing in the Act prejudices any rule of law which imposes

on the supplier a duty stricter than that imposed by Section 13 or 14 above.

It is worth noting that although Section 13 refers only to care and skill, any common law rule imposing the stricter obligation of fitness for purpose would remain effective by virtue of Section 16.

Provisions in the Design and Construct Conditions

The main provision in the D and C Conditions on materials and workmanship is clause 36 but this, of course, has to be read in conjunction with other clauses, most notably clause 5 (contract documents) and clause 8 (contractor's general obligations).

Clause 36 – materials and workmanship

Clause 36(1) is a restatement of the contractor's general obligation to design, construct and complete the works in accordance with the contract but it goes on to say 'and where not expressly provided otherwise in the Contract in accordance with appropriate standards and standard codes of practice'.

Clause 36(2) states that 'All materials and workmanship shall be of the respective kinds described in the Contract or where not so described shall be appropriate in all the circumstances'.

Thus in the absence of express standards, specifications and the like there is a fall-back provision to appropriate standards and standard codes. This is obviously useful protection to the employer in so far that his requirements are not specified in detail but nevertheless it may leave the contractor with considerable scope for choice between products of different quality.

Where materials and workmanship are expressly provided for in the contract and their respective kinds are described the obligation on the contractor in so far as clause 36 applies is apparently to do no more than comply. This is much the same as the position under the ICE Sixth edition with employer's design. However the common law and statutory obligations outlined above have to be taken into account in determining the totality of the contractor's obligation and the rules on merchantable quality and fitness for purpose have far more impact in design and construct contracting than in traditional contracting.

The question to be considered in relation to unspecified materials is who decides what is 'appropriate in all the circumstances' – the contractor or the employer? In a design and construct contract the first choice must rest with the contractor. If the employer is dissatisfied he can impose his own choice but this will probably amount to a variation under clause 51 for which he will have to pay any additional cost.

The contractor, however, must exercise his choice with caution, 'Appropriate in all the circumstances' taken together with common law and statutory obligations strongly suggests fitness for purpose. Consequently, the obligations of the contractor in respect of unspecified materials and workmanship may be greater than his obligations for specified items. The matter is complicated, however, by clause 8(2) of the Conditions.

Clause 8(2) – contractor's design responsibility

Clause 8(2)(a) states that in carrying out his design obligations, including the selection of materials to the extent they are not specified in the employer's requirements, the contractor shall exercise all reasonable skill, care and diligence.

Clause 8(2)(b) requires the contractor to accept responsibility for any design included in the employer's requirements.

On one view, the reference in clause 8(2)(a) to skill and care excludes fitness for purpose but on another view which seems more tenable the contractor is placed under an obligation to use skill and care in selecting materials which are fit for their purpose.

Thus if the contractor selects and incorporates into the works materials which are obviously unsuitable he will be in breach of clause 8(2)(a) – failing to use skill and care in selection – and clause 36(2) – failing to use appropriate materials. However if the materials prove unexpectedly to be unsuitable there may be no breach of clause 8(2)(a) although the contractor may still be in breach of clause 36(2).

The implications of the contractor accepting responsibility for the employer's design under clause 8(2)(b) are discussed in detail in Chapter 4. The intention and the mode of application of the clause is far from clear but when taken with clause 36 it is unlikely that the contractor warrants that specified materials and workmanship will be fit for purpose.

Clause 12 – unforeseen physical conditions

Clause 12 adds a further complication in that if the contractor can show that failure of materials or workmanship, whether his selection or not, is a physical condition which could not have been foreseen then the employer accepts financial responsibility for the consequences.

Clause 5 – contract documents

It is clear from clause 5 that the employer's requirements take precedence over the contractor's submission. The contractor cannot, therefore, use materials or workmanship of a lower standard than specified in those requirements. However where the employer's requirements are not precise and the contractor's submission contains ambiguities or discrepancies on standards the contractor may be able to argue that he can select the lowest standard in his submission, on the basis that it is a kind described in the contract, even if this is below the 'appropriate' standard of clause 36.

13.3 Tests and samples

There are widely differing views on the extent to which the contractor's work should be overseen and supervised in design and construct contracts.

At one end of the scale there is the philosophy of leaving the contractor to get on with it, with the employer relying on final inspection and finished quality; at the other end of the scale there is the view that regular supervision and monitoring is required on the same scale as for traditional contracting.

In the D and C Conditions the provisions for the contractor to operate a quality assurance system certainly indicate that the contractor will take a more prominent role in verifying quality than in traditional contracting and this is reflected in the drafting of clause 36.

The area where there is most likely to be difficulty in the D and C Conditions is in payment for tests and samples. This is frequently a contentious matter in the ICE Fifth and ICE Sixth edition contracts but there at least the engineer prepares all the documents and has the opportunity to specify in detail his requirements. In design and construct contracting it would be not unreasonable for an employer to assume that all costs of test and samples are included in the contractor's price. That may be the broad intention of the D and C

Conditions. Unfortunately the drafting gives the contractor a variety of opportunities to claim for extra payment.

Proposals for checking and testing

Clause 36(3)(a) requires the contractor to submit to the employer's representative proposals for:

- checking the design of the works
- checking the setting-out of the works
- testing materials and workmanship.

There is no comparable obligation on the contractor in the ICE Sixth edition but its inclusion in the D and C Conditions is entirely logical.

There is however a question raised by the opening words of clause 36(3)(a) 'Further to his obligations under clause 8(3)'. This is a reference to the quality assurance provision which itself reads 'To the extent required by the Contract the Contractor shall institute a quality assurance system'. It may be argued that if there is no requirement in the contract for quality assurance then the contractor's obligation under clause 36(3)(a) to submit proposals for checking and testing is nullified.

If such an argument were to prevail it would be unfortunate. There is a stronger practical case for the contractor to submit proposals for checking and testing if he does not have a quality assurance system than if he has. Perhaps it would be better if clause 36(3)(a) commenced 'In addition to' instead of 'Further to'.

Obligation to carry out checks and tests

Clause 36(3)(b) states the contractor's obligation to carry out checks and tests:

- as approved under sub-clause 36(3)(a), or
- as specified elsewhere in the contract.

The contractor must also carry out such further tests as the employer's representative may reasonably require.

Note that it is only 'further tests' which the employer's representative may require. There is no obligation under this clause for the contractor to carry out further checks on design or setting-out. However if it was necessary to do so the employer's representative could probably order such checks by varying the employer's requirements under clause 51.

Omitted from the D and C Conditions is the provision in clause 36(1) of the ICE Sixth edition for tests to be carried out at:

- the place of manufacture
- the site
- such other places as specified in the contract.

However, there is nothing in the Conditions to limit testing to the site so the omission may be of little significance.

Carrying out tests

Clause 36(4) is a provision taken from clause 36(1) of the ICE Sixth edition requiring the contractor to provide assistance, equipment etc for testing and to provide samples for testing. Another minor change, although possibly one of some note, is that under the ICE Sixth edition the engineer can select samples for testing; but the phrase 'as may be selected' is not included in the D and C Conditions and the employer's representative can only require samples to be supplied.

A point of more general importance is that the D and C Conditions, through clause 36(3)(b) requiring the contractor to carry out tests and clause 36(4) requiring the contractor to provide equipment etc for testing, leave no doubt that it is the contractor's responsibility to carry out tests. The ICE Sixth edition does not address this issue directly and it can be interpreted, and in practice it is frequently the case, that the contractor is required to do no more than assist the engineer in testing.

Costs of samples

Clause 36(5) provides that all samples shall be supplied by the contractor at his own cost if:

- the supply is clearly intended by the contract, or
- the supply is provided for in the contract.

Otherwise the cost of samples is to be met by the employer.

This clause is identical to clause 36(2) of the Sixth edition but there is a greater likelihood in the D and C Conditions that the employer will find himself liable for the cost of samples. If the employer's requirements are expressed only in general terms they will have nothing to say on samples; and in so far that the contractor provides the detail for the contract documents through his contractor's submission it will be to his benefit to remain silent on samples.

Tests following variations

Clause 36(6) applies to testing following variations. The clause places an obligation on the contractor to consider for each variation ordered or consented to by the employer's representative whether:

- any tests would be affected
- any tests would be appropriate.

The contractor is to inform the employer's representative without delay and proposals for amended or additional tests are to be submitted as soon as possible.

These provisions, which are not to be found in traditional ICE Conditions, relate closely to the contractor's responsibilities for design. Because the contractor's responsibility extends to variations it is essential that he should be entitled to raise testing issues. But the clause as worded goes well beyond an entitlement and indicates a duty and contractors will need to consider carefully for each and every variation how to exercise their obligations in respect of testing and the implications of failing to do so.

Contractors should have little difficulty recovering the costs of tests following variations either as a cost of the variation under clause 51 or under clause 36(7).

Costs of tests

Clause 36(7) sets out how the costs of tests are to be allocated. The clause refers only to tests and not to checks and it follows that if the contractor is to recover the costs of additional requirements for checks the basis for such recovery will have to be found elsewhere in the Conditions – such as clause 51 (variations).

The intention of the clause is much the same as that in clause 36(3) of the ICE Sixth edition. The contractor pays for tests which are clearly intended by or provided for in the contract and for additional tests the contractor pays for failures and the employer pays for passes. However allowing for the contractor's design responsibility and quality assurance obligations and the probable lump sum nature of the contract the wording of the clause leaves a number of openings for dispute.

The opening words of clause 36(7) 'Unless the Contract otherwise provides' could refer to items in a bill of quantities or allowances within a lump sum. The employer should be alert to the contractor producing contract documents which do provide such items or allowances thereby entitling the contractor to recover costs of testing which otherwise would be deemed to be included in his rates or prices.

The core of the clause is that the cost of making any test shall be borne by the contractor if such a test is:

- proposed by the contractor under clause 8(3) or clause 36(3)(a), or
- clearly intended by or provided for in the contract.

Clause 8(3) relates to the contractor's quality assurance system which as mentioned above, may or may not be obligatory. Clause 36(3)(a), which itself is in furtherance of clause 8(3), may also be of no effect in some contracts. The other side of this which contractors will need to note is that anything which the contractor proposes to do under clause 36(3)(a) is at his own cost.

The question of whether or not tests are clearly intended by or provided for in the contract is likely to be more controversial in design and construct contracting than in traditional contracting where the burden is on the engineer to ensure the contract is clear. In design and construct contracting there is no incentive on the contractor to provide for items in the contract if the contract allows for payment when they are not so provided.

Consider as an example the load testing of piles. If both the employer's requirements and the contractor's submission are silent on the matter but the employer's representative considers they are necessary how is the employer to avoid paying for the tests. The contractor can say that he never intended to carry out such tests and they are not provided for in the contract.

Clause 36(3)(b) empowers the employer's representative to order tests but the final sentence of clause 36(7) makes it clear that if he does so the contractor only pays for tests which fail and the employer pays for tests which pass.

Taken on its own the final sentence of clause 36(7) suggests that all tests, whether or not ordered by the employer's representative, are to be paid for on the above basis. The sentence commences 'If any test is carried out pursuant to sub-clause 3(b) of this Clause'. But sub-clause 3(b) refers to tests approved under clause 36(3)(a) or 'elsewhere in the Contract' and further tests ordered by the employer's representative; that is to say it covers all tests. However, such an interpretation is clearly in conflict with the remainder of clause 36(7). It would seem that the final sentence is intended to apply only to 'further tests' ordered by the employer's representative. Presumably other tests are not to be regarded as being 'pursuant to' clause 36(3)(b).

Tests following failures

The D and C Conditions do not address directly, any more than the ICE Sixth edition, the problem of testing ordered in consequence of failure in some element of the works. On grounds of prudence or safety the employer's representative may have little choice but to order testing of other work to ensure that it also is not defective. It does not seem equitable that the employer should pay for such tests even when they pass.

There is a view that under some circumstances testing under clause 36 can be regarded as a serial process, and the costs of successful follow-on tests do not automatically fall on the employer. However this is by no means a certain argument and it is disappointing that the D and C Conditions have failed to grasp this particular nettle.

13.4 *Access to site*

Clause 37, on access to site, provides that the employer's representative and any person authorised by him shall at all times have access to the works and the site.

The clause is identical to clause 37 of the ICE Sixth edition with the employer's representative substituted for the engineer. This suggests that no difference is seen between design and construct contracting and traditional contracting on the matter of access.

There is the possibility that since 'the Works' by definition in the

D and C Conditions includes design, access to design offices might be implied. But against this the references in the clause to workshops, materials, manufactured articles and machinery suggest that access to view physical work is all that is intended.

Cost of access

The clause is not restrictive on the purpose of the access and it does not seem to be open to the contractor to question why access is required. This itself may lead to dispute.

The clause places the contractor under an obligation to 'afford every facility for and every assistance in' obtaining access. Complying with this can involve the contractor in cost, delay or disruption and whereas under the ICE Sixth edition the contractor can always fall back on clause 13(3) for cost and delay arising from instructions, under the D and C Conditions the omission of clause 13 leaves the contractor with little scope for recovery for any unexpected cost.

13.5 Examination of work

The provisions in the D and C Conditions for the examination of work before covering up and the making of openings for inspection appear similar to those in the ICE Sixth edition but there is a point of major difference. In the ICE Sixth edition the contractor is not permitted to cover up work without the consent of the engineer. In the D and C Conditions, the contractor is to afford the employer's representative the opportunity to examine and measure but he is not prohibited from proceeding if the employer's representative fails to attend.

Responsibility for design

The logic of the change is obviously linked to responsibility for design. In the ICE Sixth edition the engineer is responsible for designing foundations to match the conditions encountered whereas in the D and C Conditions that responsibility rests with the contractor.

Consequently, whereas in the ICE Sixth edition clause 38 serves the dual purpose of assisting the engineer in respect of design and supervision, clause 38 in the D and C Conditions is limited in its purpose to supervision.

Clearly difficulties are likely to arise if the employer's representative under the D and C Conditions finds himself in conflict with the contractor on the state of foundations. The employer's representative may not want to concede that unforeseen conditions exist or that a change in the employer's requirements is necessary. But he has no obvious power to make the contractor comply with his wishes.

Examination before covering up

Clause 38(1) provides:

- the contractor to afford full opportunity for the employer's representative to examine and measure
- the contractor to give due notice when work is ready for examination
- the employer's representative to attend without unreasonable delay unless he considers it unnecessary and advises the contractor accordingly.

Because the contractor is not prohibited from proceeding without the consent of the employer's representative claims for delay should not arise under these Conditions.

Uncovering and making openings

Clause 38(2) gives the employer's representative power to order the contractor to uncover,or make openings in, any part of the works.

The clause provides that if the parts have been covered after compliance with sub-clause (1) and found to be in accordance with the contract the cost shall be borne by the employer, but if not by the contractor.

Again it is worth noting that the absence of prohibition from proceeding without consent makes the application of this clause different from its near identical equivalent in the ICE Sixth edition. To comply with sub-clause (1) in the D and C Conditions the contractor has to do little more than give adequate notice; he does not have to wait until he has consent. Consequently providing the contractor has given notice and allowed a reasonable time for inspection and the work is found to have been carried out in accordance with the contract, then the employer will be liable for the cost of uncovering and reinstating.

13.6 Removal of unsatisfactory work and materials

Clause 39 in the D and C Conditions is restricted solely to setting out the powers of the employer's representative to order the removal of unsatisfactory work and materials.

The clause does not, as does its ICE Sixth edition equivalent, authorise the employer to pay others to carry out the removal of unsatisfactory work and materials when the contractor is in default in compliance. Clearly such action would be incompatible with the division of responsibilities in a design and construct contract.

Nor does the clause expressly deal with failure to disapprove and the power to subsequently take action as does the ICE Sixth edition. This power, however, is not omitted from the D and C Conditions; it is simply transferred to clause 1(7)(b) – consents and approval.

Power of the employer's representative

Clause 39 gives the employer's representative power during the progress of the works to instruct in writing:

- the removal of materials which do not comply with the contract
- the replacement with materials which do comply
- the removal and replacement of work which does not comply with the contract.

Time for compliance

For the removal of materials under clause 39(a) the employer's representative has power to specify in his instruction the time or times for compliance. Surprisingly perhaps, although this is no different from the ICE Sixth edition, there is no power in this clause to specify times for the removal of work.

Default in compliance

However, under clause 65(1)(h) the contractor is in default if he fails to remove goods or materials from the site or to pull down and replace work for 14 days after receiving notice to do so. If the employer's representative certifies such default the employer may after seven days notice expel the contractor from the site.

The remedy may seem extreme and the incidence of its use is undoubtedly low but nevertheless its presence stands as a warning to contractors that the employer's representative has more than token powers under clause 39.

'During the progress of the Works'

The phrase during the progress of the works suggests that the provisions of clause 39 apply only up to the issue of the certificate of substantial completion. But since variations under clause 51 may be ordered at any time up to the end of the defects correction period this may not be the case.

It would seem that the phrase has been carried forward from the ICE Fifth edition where there was no power to order variations after completion and there was no case for the operation of clause 39 after completion.

'In the opinion of the Employer's Representative'

On matters concerning clause 39 and the removal of work and materials the opinion of the engineer and the opinion of the contractor are frequently far apart in traditional contracts. And the problem then is usually no more than disagreement on quality against specification. In design and construct contracting there is far greater scope for disagreement particularly in respect of work for which it is alleged the design does not comply with the contract.

This raises the question of what action the contractor can take if he disagrees with the opinion of the employer's representative.

The first action is to ensure that it is the opinion of the employer's representative himself and not that of an assistant acting with delegated powers. If it is the opinion of an assistant the contractor can refer the matter to the employer's representative under clause 2(6).

If that does not resolve the problem the contractor can register his dissent but he is obliged, at least in theory, to comply with the instruction and seek redress if aggrieved through conciliation and/

or arbitration. In practice where design is concerned the theory does not work without distorting the responsibilities of the parties. If the contractor is of the opinion that his design is correct he cannot be expected to pull the work down and replace it with work of some other design leaving the matter to be resolved post-contract. There needs to be a mechanism for resolving such disputes as they occur along the lines of the adjudication and expert procedures of other design and construct contracts.

13.7 *Suspension of work*

Clause 40 of the Conditions gives the employers representative the power to suspend the progress of the works. The clause is unchanged from the ICE Sixth edition except that 'employer's representative' is substituted for 'engineer' throughout.

Power to suspend

There is a legal view that without express provision in the contract for suspension of work the contractor cannot be ordered to suspend progress nor can he do so of his own choosing.

Clause 40 gives power only to the employer's representative to order a suspension. It does not give the contractor any entitlement to suspend. However it is worth noting here that the D and C Conditions have introduced in clause 64, for the first time in any standard ICE form, provision for the contractor to suspend in the event of non-payment. But that clause apart the Conditions make no provision for the contractor to suspend without an order. And if the contractor was to do so he would be potentially in breach under clause 41 – failing to proceed with due expedition and without delay; and potentially in default under clause 65 – suspending the progress of the works without due care and/or failing to proceed with due diligence. The contractor could then have his employment under the contract terminated.

Design responsibility

In traditional contracting it may be wholly logical that the contractor is given no power to suspend. His task is to construct what the engineer has designed. However in traditional contracting if the engineer wants to halt work whilst he re-assesses his design he has the power to do so.

It may be a matter of concern that in the D and C Conditions no similar power is given to the contractor to accompany his responsibility for design. If the contractor wants to halt work to re-assess his design it would seem that his best course would be to persuade the employer's representative that he should order a suspension 'for the proper construction and completion' of the works.

Reasons for suspension

Clause 40 is not concerned with reasons for suspension except in connection with the contractor's right to payment of extra cost and extension of time for completion. The clause gives the employer's representative discretionary power to suspend on whatever grounds he considers necessary. And the wording appears to be wide enough to allow the employer's representative to act on the instructions of the employer as well as on his own professional judgment.

Reasons for suspending progress will probably fall under one of the following broad headings:

- change of mind by the employer
- possession and access problems
- unexpected restrictions imposed on the works
- pre-planned closures
- unforeseen conditions encountered
- adverse weather
- safety
- unsatisfactory performance by the contractor.

Temporary lack of funds by the employer might occasionally be added to the list although there is a case for saying this may be an abuse of clause 40. From the location of the clause in the Conditions and its wording it is clearly intended principally for practical matters relating to when and how the works are constructed and not to financial matters.

Operation of clause 40

The procedural arrangements of clause 40 are as follows:

- The employer's representative gives a written order to suspend the progress of the works – or any part thereof – for such time or times – and in such manner – as the employer's representative considers necessary.
- During the period of the suspension the contractor is to protect and secure any work, so far as is necessary in the opinion of the employer's representative.
- The contractor is entitled to be paid, subject to clause 53 which relates to notice, the extra cost of giving effect to the employer's representative's order and instructions unless the suspension is:
  - provided for in the contract
  - necessary by reason of weather conditions
  - necessary because of contractor's default
  - necessary for proper execution of work
  - necessary for safety.

  Profit is to be added to cost in respect of any additional permanent or temporary work.
- The contractor is entitled to an extension of time for completion for delay caused by a suspension, unless the suspension is:
  - otherwise provided for in the contract, or
  - necessary by reason of some default of the contractor.

Shared responsibility

Clause 40(1)(c) recognises that a suspension for proper execution or safety could arise from some default of the employer's representative or employer or could be an excepted risk under clause 20(2). Of these, use by the employer is the most obvious. The exception to the contractor's right to payment is modified accordingly in such circumstances.

Provided for in the contract

The exception to the contractor's right to payment in clause 40(1)(a) is expressed as 'otherwise provided for in the Contract'.

It is suggested that this relates only to specific close-down periods which are mentioned in the contract documents. It is not thought to be a reference to the provisions for suspension in clause 12 (unforeseen physical conditions) or clause 64 (employer's failure to pay).

Weather conditions

Suspension for weather conditions is usually given at the contractor's request, since although there is no recovery of cost, there is often entitlement to extension of time. The position when the employer's representative suspends for weather conditions against the contractor's wishes is potentially contentious. The employer's representative will probably have in mind that temperatures are too low or too high for compliance with specifications or that working in wet weather or snow will render materials unsuitable. The contractor will often contend that he proceeds at his own risk and the employer's representative has no right to interfere. However, by clause 40(1) the employer's representative has a right to interfere and it is suggested he has a duty to the employer to prevent needless destruction of materials where the costs of replacement fall on the employer.

In design and construct contracting, particularly with lump sum contracts, such costs are more likely to fall on the contractor and accordingly there may be a more realistic acceptance by contractors that suspensions in adverse conditions are to the good of both parties.

Contractor's default

The exception to payment and entitlement to extension of time in clause 40 expressed as 'some default on the part of the Contractor' could cover more than just physical circumstances. It could extend to defaults in design or administration. One example might be breakdown of the contractor's quality assurance system. Another could be failure by the contractor to submit designs and drawings for consent.

There is certainly more scope for default in design and construct contracting than in traditional contracting and it may well be that suspensions for default in the D and C Conditions will take on more significance than is the case with the ICE Fifth and ICE Sixth editions.

Suspension longer than three months

Clause 40(2) deals with a suspension which has not been lifted within three months. Unless the suspension is either:

- provided for in the contract, or
- necessary by reason of contractor's default.

The contractor may give written notice requiring permission to proceed within 28 days and if the employer's representative does not grant such permission the contractor may, by further written notice:

- treat suspension of part of the works as a variation under clause 51, or
- treat suspension of the whole of the works as abandonment of the contract by the employer.

Omission variations

There is nothing in clause 40(2) in its references to variations under clause 51 which distinguishes it from clause 40(2) of the ICE Sixth edition. But clause 51 of the D and C Conditions is very different from clause 51 of the ICE Sixth edition and most particularly it is different in respect of omission variations and their valuation. See Chapter 16 for further comment on this.

Under the ICE Sixth edition if there is an omission variation it is up to the engineer to give instructions on the design and the practical implications of the change. But under the D and C Conditions the responsibility for dealing with the change would seem to fall on the contractor.

And as to valuations, there is no certainty that the D and C Conditions provide for deduction from a lump sum price for an omission variation.

13.8 Materials and contractor's equipment

Clause 54 of the D and C Conditions deals principally with the vesting of goods and materials not on site where the contractor is seeking payment under clause 60. For this purpose the clause corresponds closely to clause 54 of the ICE Sixth edition.

However, because clause 53 of the D and C Conditions has been allocated to additional payments and claims procedures, part of clause 53 of the ICE Sixth edition, dealing with contractor's equipment, is to be found in clause 54 of the D and C Conditions.

Contractor's equipment

Clause 53 of the ICE Sixth edition is divided into three sub-clauses:

- clause 53(1) which has two provisions:
  - the contractor's equipment etc is deemed to be the property of the employer when on site
  - the contractor's equipment etc. not to be removed from site without consent
- clause 53(2) – the employer not liable for loss or damage to the contractor's equipment
- clause 53(3) – the employer can dispose of contractor's equipment left on site.

In the D and C Conditions the contractor's equipment is not deemed to be the property of the employer. Consequently there is no equivalent to the first part of clause 53(1) or clause 53(2) of the ICE Sixth edition. All that is retained of clause 53(1) is the second part on non-removal of the contractor's equipment without consent. This forms clause 54(1) of the D and C Conditions.

The provision in clause 53(3) of the ICE Sixth edition on disposal of contractor's equipment left on site is transferred to clause 33(2) of the D and C Conditions without any significant change of the wording. In effect if the contractor fails to remove his equipment within a reasonable time after completion the employer can dispose of the same and retain the costs of disposal before paying the balance to the contractor.

Vesting of goods and materials not on site

Clauses 54(2) to 54(7) of the D and C Conditions detail the arrangements for vesting ownership in the employer before the contractor can be paid for goods and materials not on site.

Clause 54(2) provides that with a view to securing payment the contractor may, and shall if the employer's representative so directs, transfer ownership of goods and materials not on site where such goods and materials are:

- listed in the form of tender, or
- agreed between the contractor and employer before delivery, or
- manufactured and ready for incorporation in the works, and

- are the property of the contractor or the contract of supply makes provision for them to pass unconditionally to the contractor.

Action by contractor

As evidence of ownership of off-site goods and materials, the contractor is required by clause 54(3) to take, or cause his supplier to take action as follows:

- providing the employer's representative with documentary evidence that property has vested in the contractor, marking and identifying goods and materials to show:
  (a) their destination is the site
  (b) they are the property of the employer
  (c) to whose order they are held
- setting aside and storing to the satisfaction of the employer's representative
- sending the employer's representative a list and schedule of values
- inviting the employer's representative to inspect.

Employer's representatives should ensure that these requirements are strictly fulfilled because of the inevitability of conflicting claims to ownership if there is a business failure somewhere along the lines of supply or the contract chain.

Vesting in employer

Clause 54(4) recognises that the employer obtains full ownership under clause 54.

It, therefore, makes necessary provision for:

- rejection of goods and materials not in accordance with the contract
- ownership to revest in the contractor upon rejection
- the contractor to be responsible for loss or damage
- the contractor to take our additional insurance.

Lien on goods or materials

Clause 54(5) is intended to prevent the contractor, sub-contractor or any other person holding the goods or materials which have vested in the employer against sums they allege to be due.

This may be legally effective against the contractor or his legal successors but it is of limited practical value against third parties.

Delivery of vested goods and materials

Clause 54(6) gives the employer the right to obtain delivery of, or take possession of, goods or materials after cessation of the contractor's employment under clause 63, 64, 65 or otherwise.

Incorporation into sub-contracts

Clause 54(7) is intended to ensure that the contractor's rights against sub-contractors are no less than the employer's against the contractor so that in the event of default in the contractual chain the employer's title is not lost.

Chapter 14

Commencement, possession, time and delays

14.1 *Introduction*

The clauses in the D and C Conditions relating to commencement, possession, time for completion and damages for late completion are essentially the same as in the ICE Sixth edition. A few minor changes in administrative detail have been made but apart from the omission of increased quantities as a relevant event for extension of time there is nothing in the Conditions to indicate that a design and construct contract is any different from a traditional contract on matters of commencement, possession and time.

Thus, unlike The Department of Transport design and construct contract, no distinction is made between the commencement of design work and the commencement of work on site. Clause 42 still permits the contract to prescribe the order in which the works shall be constructed – an arrangement which does not sit easily with the contractor's responsibility for both design and construction. And no specific provision is made in clause 44 for delays in the design and approval process although by clause 26 the employer warrants to obtain all necessary planning permissions in due time.

It may well be that for most contracts let under the D and C Conditions these will be matters of little consequence but there is the possibility that in a minority of contracts the efficacy of the Conditions will be severely tested.

14.2 *Commencement*

The D and C Conditions use the phrase 'commencement date' instead of 'works commencement date' as used in the ICE Sixth edition.

Commencement date

The commencement date is defined in clause 1(1)(n) as the date defined in clause 41(1). Clause 41(1) states that the commencement date shall be:

- the date specified in the appendix, or if no date is specified
- such other date as may be agreed, or in default of agreement
- 28 days after the award of the contract.

In effect this means that if no date is specified in the appendix then the contractor is entitled to a start date 28 days after the award of contract although he may, but is not obliged to, agree to some other date. And likewise, if no date is specified, the employer is entitled to insist that the contractor should commence 28 days after the award.

Note that because provision for specifying the commencement date appears only in part 1 of the appendix, and not in part 2 which is filled in by the contractor, only the employer can specify the commencement date. In practice, however, there may be sound reasons for seeking the views of tenderers on the commencement date. Such consultation could result in the specification of a date which although not in the appendix could be made contractually binding by pre-award exchange of letters.

Note also a subtle change of wording between the D and C Conditions and the ICE Sixth edition in connection with the 28 day period. The D and C Conditions are definitive – the commencement date shall be 28 days after the award of the contract. In the ICE Sixth edition the wording is 'a date within 28 days of the award' – with the effect that the 28 day period acts only as a long stop and the engineer is empowered, potentially unfairly, to fix an earlier date.

'Other date as may be agreed'

The question will inevitably arise – can the employer avoid both the specified date and the 28 day date by writing 'to be agreed' in the appendix?

The answer, it is suggested, is that he cannot. The law does not favour agreements to agree because of the uncertainty which follows from failure to agree. Either one party accepts the will of the other or there is deadlock.

The purpose of the provision for agreement of the commencement date in clause 41(1) is probably no more than to allow advancement or deferment of the 28 day date by genuine agreement of the parties.

Delay in the commencement date

Under traditional ICE contracts prior to the ICE Sixth edition contractors often complained of unreasonable delay in fixing the commencement date with the effect that they suffered extra costs. The D and C Conditions should avoid such claims but there remains the possibility that for one reason or another the employer will not be in a position to permit commencement by the specified date or the 28 day date and will be obliged to seek deferment.

In such circumstances the contractor can pursue a claim for breach of contract if loss or damage results. This will be a common law claim since the contract does not provide for this particular breach.

Effect of the commencement date

The effect of the commencement date is twofold. It fixes the date on which the contractor's obligations to proceed commence; and it fixes the date for completion of the works in that it is the date from which the time for completion is calculated. It is, of course, essential in a contract which specifies a time for completion rather than a date for completion that there is a reference point for commencement.

It is an interesting point that in the D and C Conditions unless the commencement date is specified in the appendix it may never be fixed in writing. It may be agreed orally or it may be calculated by reference to the 28 day period. The parties should however avoid oral agreements.

In the very old case of *Kemp* v. *Rose* (1858) where the date for commencement was omitted from a written contract the court declined, in the face of conflicting oral evidence, to set a date.

14.3 Start of the works

Clause 41(2) requires the contractor to start the works on or as soon as is reasonably practicable after the commencement date. Thereafter the contractor is to proceed with due expedition and without delay in accordance with the contract.

This is identical to clause 41(2) of the ICE Sixth edition and although it appears to make no allowance for any design which is necessary in advance of commencement the explanation lies in the definition of 'the Works'. By clause 1(1)(m) the 'Works' means the permanent and temporary works and the design of both. Consequently in starting design the contractor is starting the works.

As soon as is reasonably practicable

In the D and C Conditions the phrase 'as soon as is reasonably practicable' will involve different criteria than it does in the ICE Sixth edition and other traditional contracts.

In such contracts the phrase is usually taken to relate principally to the contractor's mobilisation period and the delivery times on items of plant and materials. However in the D and C Conditions the contractor's design responsibility will not only be relevant but possibly the most significant factor in assessing what is as soon as reasonably practicable.

Thus clause 8(2)(b) requires the contractor to check any design undertaken by or on behalf of the employer and to accept responsibility for such design. Whether checking another person's design is to be regarded as designing the works is a point for debate but the prudent contractor faced with a near completed employer's design will certainly not want to start the works until he is satisfied on the integrity of the design.

Failure to commence

Failure by the contractor to commence the works in accordance with clause 41 is a default under clause 65 which can lead to determination of the contractor's employment.

It is, however, first necessary for the employer's representative to certify in writing that the default has occurred.

Whilst this may be a straightforward matter on a traditional contract where the contractor has failed to appear on the site it will be anything but straightforward in the D and C Conditions where design work in far away offices is concerned.

Due expedition and without delay

The same comments apply to the requirement for the contractor to proceed with due expedition and without delay. Assessing these matters for work on site is one thing; assessing them for design is another.

Although failure by the contractor to proceed with due expedition and without delay is a breach of contract it is not fully clear whether or not the employer has a remedy other than damages for late completion. Clause 65 includes the default of failing to proceed with due diligence as ground for determination but it is arguable that diligence relates to correctness and that it does not extend to expedition.

In accordance with the contract

The closing phrase of clause 41(2) 'in accordance with the Contract' is sometimes interpreted in traditional ICE contracts as placing an obligation on the contractor to comply with his programme.

Whilst it is doubtful if there can be any such obligation in respect of the clause 14 programme the position may be different for a tender programme listed as a contract document.

The phrase may relate to prescriptions on possession of the site and access given under clause 42 and it may link in with clause 46 which deals with the expedition of progress and accelerated completion. However, an alternative explanation is that the phrase is intended to place assessment of 'due expedition and without delay' in the context of the time allowed for completion in the contract. This would certainly be consistent with provisions for damages which apply only to late completion.

It is possible, therefore, that the phrase 'in accordance with the Contract' does not add to the contractor's obligation to proceed with due diligence and without delay but it simply qualifies that obligation so that instead of being absolute the obligation is secondary to finishing on time.

14.4 Possession of the site and access

Clause 42, which is taken almost word for word from the ICE Sixth edition, sets out the employer's obligations in giving possession of the site and providing access; and it details the contractor's remedy in the event of default.

By the standards of construction contracts outside the ICE family of forms the clause is unusually elaborate but it is a feature of ICE conditions that the contractor is not entitled to full possession of the site at commencement nor is he entitled to exclusive possession. Since an argument could be advanced in the absence of express provisions that the contractor has such rights, the detail of the clause serves generally to good effect.

Meaning of possession

Possession of the site under the contract does not mean that the contractor is literally the party in possession. The word 'possession' is used for convenience and it is not intended to confer on the contractor the full range of legal rights and liabilities which accompany possession by ownership.

Possession given to the contractor is more in the nature of a licence to occupy the site. Thus in *H W Neville (Sunblest) Ltd* v. *William Press & Sons Ltd* (1981) it was said:

> 'Although [the contract] uses the word 'possession' what it really conferred on William Press was the licence to occupy the site up to the date of completion.'

And in *Surrey Heath Borough Council* v. *Lovell Construction* (1988) the following passage from Hudson on Building Contracts (10th edition) was accepted as an accurate statement of the law:

> 'In the absence of express provisions to the contrary, the contractor in ordinary building contracts for the execution of the works upon the land of another has merely a licence to enter upon the land to carry out the work. Notwithstanding that contractually he may be entitled to a considerable degree of exclusive possession of the site for the purpose of carrying out the works, such a licence may be revoked by his employer at any time, and thereafter the contractor's right to re-enter upon the site of the works would be lost. The revocation, however, if not legally justified, will render the employer liable to the contractor for damages for breach of contract, but subject to this the contractor has no legally enforceable right to remain in possession of the site against the wishes of the employer.'

Clause 42(1) – prescriptions on possession and access

Clause 42(1) states that the contract may prescribe:

- the extent of portions of the site which the contractor is to be given possession of from time to time

- the order in which such portions shall be made available to the contractor
- the availability and nature of the access to be provided by the employer
- the order in which the works shall be constructed.

There is no entry in the appendix for these prescriptions and in the ICE Sixth edition they will be found in the drawings, specification or special conditions. But they will, of course, be prescriptions imposed by the employer.

In the D and C Conditions the contract covers both the employer's requirements and the contractor's submission. It would seem to be open, therefore, for the contractor to impose prescriptions and providing they are not in conflict with the employer's requirements they would appear to be contractually binding.

The implications of this could be quite startling. The contractor could, for example, specify in his submission the nature of the access to be provided by the employer.

Perhaps the opening words of clause 42(1) 'The Contract may prescribe' are of wider application than their intention and should be redrafted.

Order of construction

There are clear dangers in fixing the order of construction by contractual requirements because any directions on how the contractor is to operate can rebound on the employer as claims. The employer starts with the basic obligation not to impede the contractor in the performance of his obligation to complete the works on time. But by prescriptions, or directions, the employer can end up with subsidiary obligations.

The case of *Yorkshire Water Authority* v. *Sir Alfred McAlpine & Son (Northern) Ltd* (1985) illustrates the point. In that case the contractor submitted with his tender, as instructed, a method statement, showing he had taken note of certain specified phasing requirements providing for the construction of the works upstream. The formal contract agreement incorporated the method statement.

The contractor maintained that in the event it was impossible to work upstream and after some delay work proceeded downstream. The contractor then sought a variation order under clause 51(1). The court held:

- the incorporation of the method statement into the contract imposed an obligation on the contractor to follow it so far as it was legal or physically possible to do so;
- the method statement, therefore, became a specified method of construction and the contractor was entitled to a variation order and payment accordingly.

Clause 42(2) – possession and access

The obligations of the employer in giving possession and access are set out in clause 42(2). These can be summarised as follows:

- to give possession on the commencement date of so much of the site as required to enable the contractor to commence and proceed;
- thereafter to give possession of such further portions of the site as required by the accepted clause 14 programme;
- to give access on the commencement date to enable the contractor to commence and proceed;
- thereafter to give such further access as necessary to enable the contractor to proceed with due despatch.

It can be questioned why the obligations on access are not drafted to match those on possession. At face value it would seem that if the contractor gets ahead of his clause 14 programme he is not entitled to further possession but he is entitled to further access.

The explanation is probably that 'further access' is intended to apply only to 'further portions'. But in a situation where full possession is given at commencement and only the availability of access is phased then 'further access' cannot be so construed.

Failure to give possession

Clause 42(3) provides for the contractor to be given an extension of time and recovery of additional cost for any delay or cost caused by the employer's default in giving possession.

As in the ICE Sixth edition the clause expresses the contractor's entitlement more positively than other claim clauses in the Conditions; and it can be read as placing a duty on the employer's representative to act independently of an application from the contractor. The clause even requires the employer's representative to notify both the parties of any extension of time awarded.

This approach may be because of the nature of the breach in failing to give possession. At common law such a breach could be considered fundamental thereby entitling the contractor to determine. But since this remedy is not given to the contractor under the Conditions perhaps it is considered necessary to ensure that the contractor is automatically compensated for the employer's default – so avoiding the possibility of common law determination.

Failure to give access

Clause 42(3) refers only to failure to give possession. The clause says nothing about failure to provide access.

There is, of course, a difference between the two obligations. The employer is wholly bound to give possession and the contract can only prescribe the extent and order in which portions of the site are to be given. The employer is not bound to provide access. In clause 11(2)(c) the contractor is to satisfy himself on the means of access he requires.

Nevertheless if the contract prescribes for access to be provided by the employer it is surely a promise which amounts to a contractual obligation. So although the contract is silent on the point the employer is certainly liable for damages for breach.

In the event of delay being caused the contractor might well be able to claim time to be at large since there is no provision in the Conditions for extending time for this occurrence.

Clause 42(4) – access provided by the contractor

Clause 42(4) states, presumably for the avoidance of doubt, that the contractor shall bear all costs and charges for any access additional to that provided by the employer. The obligation would seem to arise in any event under clause 11 which refers specifically to access.

Clause 42(4) also states that the contractor shall provide at his own cost any additional facilities outside the site he requires for the purposes of the works. As with the ICE Sixth edition this provision does not sit easily with clause 26(2) which states that the employer will pay the rates and taxes on temporary structures outside the site used exclusively for the works and on other structures used temporarily and exclusively for the purposes of the works. It would appear that the cost of 'providing' is not to be taken as paying rates and taxes.

14.5 *Time for completion*

Clause 43 provides confirmation of the contractor's obligation to complete the works within the time stated in the appendix or as extended.

In the form of tender and the form of agreement the contractor undertakes only to complete in conformity with the provisions of the contract.

Sectional completion

The obligation of completion in clause 43 extends to any section for which a particular time is stated in the appendix. The scheme is that there should be stated in the appendix either:

- a time for completion of the whole of the works, or
- times for various sections with the 'remainder' of the works forming a final section.

All times for completion of sections run from the single commencement date and are not intermediate periods.

Contractor's own times

The Conditions make provision for the practice of allowing the contractor to state his own times for completion. They do this by a two-part appendix to the form of tender. The employer can either insert required times for completion in part 1 or allow the contractor to insert his own times in part 2.

The advantages of giving the contractor freedom are firstly, tenders can be compared on time as well as price, and secondly, disputes on shortened programmes should largely be avoided.

Completion of phases (or parts)

Clause 43 places no obligation on the contractor to complete phases or parts of the works by times which may be specified in the specification, bills or drawings, unless those phases or parts are designated as sections and are properly included in the appendix.

Correspondingly, there is no obligation on the employer to take possession of a phase or part not so designated unless it has been occupied or put into use. By clause 48 the employer is however obliged to accept any section for which the employer's representative has issued a certificate of substantial completion.

Damages for late completion of phases

Because the Conditions do not oblige the contractor to complete phases or parts, not designated as sections, within specified times, the provisions for liquidated damages do not apply, and cannot be applied, to late completion of phases or parts.

In *Bruno Zornow (Builders) Ltd* v. *Beechcroft Developments Ltd* (1990) a contract was negotiated for a housing development on the basis of a first tier tender which showed a detailed programme to complete in 16 months and second stage agreements for completion of the work in two overlapping phases. The architect calculated liquidated damages of £40 000 based on the stipulated rate of £200 per week per block from the date shown on the original works programme and the contractor sued for the return of this amount. It was held:

- the contract did not incorporate documents which specified dates for sectional completion but only phased provisions for the transfer of possession;
- a claim for liquidated damages could only be made in respect of failure to meet specified completion dates and not failure to meet transfer of possession dates – which operated on a consent basis;
- no term would be implied for any sectional dates for completion.

A similar situation arose in *Turner* v. *Mathind* (1986) where there was a clear requirement in the bills for phased completion but the sectional completion supplement was not used and the appendix contained only a rate for liquidated damages for late completion of the whole of the works. It was held that it was not appropriate that the employer should either pro-rata the stipulated damages to the number of phases or apply the stipulated rate to each phase.

Completion within the time allowed

Both clause 43 and the appendix state the contractor's obligation to be to complete 'within' the time allowed. The contractor is therefore entitled to finish early.

14.6 Extension of time for completion

Clause 44 deals with extensions of time for completion. It specifies the events which entitle the contractor to an extension and sets out the procedures to be followed by the contractor and the employer's representative.

The only changes from the ICE Sixth edition are that the employer's representative replaces the engineer and increased quantities are omitted from the specified events.

Purposes of extension provisions

The contractor is under a strict duty to complete on time except to the extent that he is prevented from doing so by the employer or is given relief by the express provisions of the contract. The effect of extending time is to maintain the contractor's obligation to complete within a defined time and failure by the contractor to do so leaves him liable to damages, either liquidated or general, according to the terms of the contract. In the absence of extension provisions, time is put at large by prevention and the contractor's obligation is to complete within a reasonable time. The contractor's liability can then only be for general damages; but first it must be proved that the contractor has failed to complete within a reasonable time.

Extension of time clauses, therefore, have various purposes:

- to retain a defined time for completion
- to preserve the employer's right to liquidated damages against acts of prevention
- to give the contractor relief from his strict duty to complete on time in respect of delays caused by specified neutral events.

Relevant events

The events which entitle the contractor to extensions of time are commonly called 'relevant events'. Clause 44(1) specifies the events which apply to these Conditions. Other contracts have more or fewer relevant events according to their policy on risk sharing and the extent to which reliance is placed on catch-all phraseology such as 'any default of the employer' or 'circumstances outside the contractor's control'.

In clause 44 the relevant events are:

- any variation ordered under clause 51(1)
- any cause of delay referred to in the Conditions
- exceptional adverse weather conditions
- other special circumstances of any kind whatsoever.

Variations

The ordering of variations is probably the most commonplace form of prevention by the employer of the contractor's obligation to complete on time – at least as far as construction contracts are concerned. In design and construct contracts however there should be fewer formal variations than in traditional contracting – providing the employer's requirements are competently drafted – since the responsibility for detailing rests with the contractor. And in traditional contracting many of the variations claimed by contractors are not so much additional works but changes in construction details.

The variation clause of the ICE Sixth edition with its extensive list of items which constitute a variation is an encouragement to such claims but the position in the D and C Conditions is quite different. The only variations referred to in clause 51(1) are variations to the employer's requirements.

The implications of this narrow definition of variations are discussed in detail in Chapter 16 but the point of interest here is whether it significantly reduces the contractor's entitlement to extension of time. It may well do so.

Since there is no clause 13 in the D and C Conditions to link clause 51 with instructions and impossibility, the application of clause 51 appears to be solely to ordered variations. For example it is difficult to see how the contractor in the *Yorkshire Water* case referred to earlier in this chapter would have won his right to a variation if the contract had been under the D and C Conditions.

Delays referred to in the Conditions

The delays referred to in the Conditions are most obviously those in clauses with specific reference to clause 44. There can be delays under other clauses without such reference – e.g. clause 32 (fossils); clause 20 (excepted risks) – and here it is suggested the extension would have to be given under 'any special circumstances'. However, as discussed under 'special circumstances' below, it may well be that some delays which can readily be contemplated and which are attributable to the employer are not covered by either category.

The delays in the Conditions with specific reference to clause 44 are listed below alongside corresponding provisions in the ICE Sixth edition for comparison.

The D and C Conditions	**The ICE Sixth edition**
• clause 5(1)(c)(ii) delay caused by instructions resolving ambiguities or discrepancies in the employer's requirements	• clause 5 no direct link with clause 44; route through clause 13
• clause 6(1)(b) failure of the employer's representative to issue information within a reasonable period	• clause 7(4)(a) failure of the engineer to issue drawings etc within a reasonable period
• clause 6(2)(d) failure of the employer's representative to issue consent or notice of withholding consent of contractor's designs and drawings	• clause 7(6) does not mention this delay
• clause 8(3) failure of the employer's representative to approve or reject the contractor's quality assurance plan	• no equivalent
• clause 12(6)(b) delay caused by unforeseen physical conditions	• clause 12(6) delay caused by unforeseen physical conditions

The D and C Conditions	The ICE Sixth edition
• no equivalent	• clause 13(3) delay caused by engineer's instructions
• no direct equivalent but see clause 6(2)(d)	• clause 14(8) delay in engineer's consent to contractor's methods or caused by engineer's requirements
• no equivalent	• clause 27(6) delays attributable to street works variations
• clause 31(2) delay caused by providing facilities for other contractors	• clause 31(2) delay caused by providing facilities for other contractors
• clause 40(1) delay caused by suspension of work	• clause 40(1) delay caused by suspension of work
• clause 42(3) failure of the employer to give possession of the site	• clause 42(3) failure of the employer to give possession of the site
• clause 64(3) delay caused by suspending or reducing work rate in consequence of failure to certify or pay amount due to the contractor.	• no equivalent.

Exceptional adverse weather

Adverse weather of itself does not give any grounds for non-performance of contractual obligations. Unless there are provisions in the contract offering relief, the contractor is deemed to have taken all risks from weather. Some recent contracts including the Department of Transport's design and construct conditions and the Department of the Environment's GC/Works/1 – edition 3 leave the risk with the contractor.

The extent to which adverse weather applies as a relevant event depends on the wording of the contract. The D and C Conditions use the phrase 'exceptional adverse weather' conditions but it is not just the phrase which has to be considered but also its context; there has to be delay, not just exceptionally adverse weather. This may seem theoretical but it is common for contractors to apply for extensions of time on the grounds that the weather has been worse than average and sight can become lost of the need for proof of delay. The practice of obtaining local weather records and comparing them on a year to year basis, or on a particular year against average, may show that the weather has been exceptional but it says nothing about delay.

The point came up in the case of *Walter Lawrence & Son Ltd* v. *Commercial Union Properties (UK) Ltd* (1984) where a contractor was suing for return of amounts deducted as liquidated damages. It was held that:

> '... When considering an extension of time under clause 23(b) of JCT 63, on the ground of 'exceptionally inclement weather' the correct test for the architect to apply is whether the weather itself was 'exceptionally inclement' so as to give rise to delay and not whether the amount of time lost by the inclement weather was exceptional ...'

Another matter of interest arose in the *Walter Lawrence* case in respect of the time at which the weather should be assessed. The architect in correspondence had said: 'It is our view that we can only take into account weather conditions prevailing when the works were programmed to be put in hand, not when the works were actually carried out'.

The contractor refuted this and claimed that his progress relative to programme was not relevant to his entitlement to an extension. It was held that the effect of the exceptionally inclement weather is to be assessed at the time when the works are actually carried out and not when they were programmed to be carried out even if the contractor is in delay.

Perhaps is should be added however that the judge in the *Walter Lawrence* case drew a distinction between delays which occur during the original or extended time for completion and delays after the due date when the contractor is in culpable delay.

'Other special circumstances'

There are differing views on what can be included within the scope of 'other special circumstances'. The argument has been going on for years with the ICE Fifth and ICE Sixth editions.

On one view the wording of the phrase is so wide – 'other special circumstances of any kind whatsoever which may occur' – that there are no exclusions. The other view is that catch-all phrases cannot include for unspecified breaches of contract by the employer. In *Fernbrook Trading Co. Ltd* v. *Taggart* (1979) it was held that the words 'any special circumstances of any kind whatsoever' were not wide enough to empower the engineer to extend time for delays caused by the employer's breaches of contract in making late payments on interim certificates.

The matter is of considerable importance because if the employer causes delay and there is no provision for extension, then the liquidated damages provisions fall and the employer can only recover such general damages as he can prove for the contractor's failure to complete within a reasonable time.

Such delays within the D and C Conditions could include:

- delay caused by rectifying damage arising from use or occupation of the works by the employer – clause 20(2)(a)
- delay by the employer in obtaining planning consent – clause 26(3)(c)
- delay caused by uncovering work or making openings where the contractor is not at fault – clause 38(2).

Numerous similar delays can be identified and whilst in the ICE Sixth edition many can be brought within the scope of clause 44 via clause 13 on engineer's instructions there is no such route in the D and C Conditions which have no clause 13.

Employers will have to decide whether they are prepared to risk the wording on other special circumstances as it stands in clause 44 or whether to amend by adding reference to employer's default.

Leaving aside the above argument there is little consensus on what situations should qualify as other special circumstances. A delay caused by the discovery of antiquities (clause 32) would not be controversial but delays caused by damage to the works, industrial disputes, non-availability of materials and the like do not automatically qualify and might well be rejected. There is no consistency of approach between engineers, employer's representatives or arbitrators.

Contractor's application

Clause 44(1) provides that should the contractor consider that any of the relevant events entitles him to an extension of time, he shall: 'within 28 days after the cause of delay has arisen or as soon thereafter as is reasonable' deliver to the employer's representative full and detailed particulars in justification of the extension claimed.

The clause does not require the contractor to give notice of delay unless he is seeking an extension – which is a curious omission in the light of clause 44(2) – assessment of delay. This suggests that it is open to the contractor to claim an extension of time whether or not he expects to overrun the specified time for completion.

Failure by the contractor to make an application under clause 44(1) does not automatically disentitle him to an extension. The employer's representative is required to assess the contractor's entitlement at the date for completion and on the issue of the certificate of substantial completion whether or not the contractor makes an application.

Where the contractor will be at risk is in making a late application, or no application at all, in respect of neutral events such as adverse weather. He will similarly be at risk in gaining his full entitlement for events where, by his conduct, he has prevented the employer's representative from investigating the delay.

Clause 44(2) – assessment of delay

The procedure in clause 44(2) for assessment of delay was introduced first into the ICE Sixth edition. It requires the employer's representative to respond to the contractor's application for extension of time by assessing the delay as a result of the alleged cause before going on to consider whether or not any extension is due.

The intention of this is apparently:

- to establish and place on record details of delays for availability in subsequent disputes; and
- to avoid the granting of extensions of time when they are not strictly necessary in relation to the time for completion.

The procedure does not appear to act as a strict condition precedent to granting extensions of time. Although it is clearly intended to operate prior to the granting of interim extensions under clause 44(3) it is doubtful if it has any application to assessments under clauses 44(4) and 44(5). It is unlikely that extensions granted without operation of the procedure would be invalidated and perhaps it should be seen as supplementary rather than integral to the rules for extension of time.

The detail of clause 44(2) is less than precise. No time limits are imposed for action, unlike clause 44(3) which requires extensions to be granted forthwith and clauses 44(4) and 44(5) which have 14 day limits.

Moreover, although the employer's representative can make an assessment in the absence of a claim he can only make an assessment of the delay caused by matters defined as relevant events for extensions. So he is not entitled, although it might be useful, to issue a notice under this clause where it is apparent that the contractor has been delayed by circumstances within his control which give no entitlement to an extension.

Clause 44(3) – interim extension of time

There is much to be said for an arrangement which entitles the contractor to be granted extensions of time in time to use them – rather than granting them after completion when costs may have been incurred on acceleration measures. This is why provisions for interim extensions were introduced in the ICE Fifth edition in 1973 and why they remain in the D and C Conditions.

Clause 44(3) provides that an interim extension should be granted 'forthwith' should the employer's representative consider that the delay suffered fairly entitles the contractor to such extension. Failure by the employer's representative to operate this provision may leave the employer open to claims for constructive acceleration.

If the contractor has made a claim but the employer's representative does not consider any extension to be due, the contractor is to be informed 'without delay'. Surprisingly this need not be in writing.

Clause 44(4) – assessment at the date for completion

Clause 44(4) places a firm duty on the employer's representative to consider the contractor's entitlement to extension of time not later than 14 days after the due date or extended date for completion of the works or any section. This must be done whether or not the contractor has made an application.

The employer's representative is to consider all the circumstances known to him at the time – that is to say he is not confined to those matters notified to him by the contractor. See below for comment on reductions.

The procedure under clause 44(4) is clearly intended to operate on a repeat basis and it must be put into effect after each extended completion date has expired.

Note that, if the employer's representative does not consider the contractor entitled to an extension of time, he is to notify both the employer and the contractor. This is presumably so that the employer can consider deducting liquidated damages under clause 47.

Clause 44(5) – final determination of extension

Clause 44(5) requires the employer's representative to make a final review of 'all the circumstances of the kind referred to in sub-clause (1)' and to finally determine and certify the 'overall' extension due. This is to be undertaken within 14 days of the issue of the certificate of substantial completion.

The final review is to consider the 'overall' extension due. It is not therefore a matter of granting incremental extensions as in clauses 44(3) and 44(4) but rather a matter of fixing a final date for completion. This is reflected in the requirement for the employer to receive a copy of the certificate issued to the contractor.

The change of wording from the employer's representative considering 'all the circumstances known to him' in clause 44(3) to 'all the circumstances of the kind referred to in sub-clause (1)' is interesting. It may not indicate a great deal but perhaps it illustrates that the clause 44(5) review should be undertaken on an analytical basis rather than on a response basis.

A minor oddity in clause 44(5) is that it requires the employer's representative to finally certify within 14 days of the issue of the certificate of substantial completion. But under clause 44(1) the contractor has 28 days from an event to make an application for an extension. This could result in the final review preceding notification of delays occurring in the fortnight before completion.

A further small point to note is that unlike the ICE Sixth edition with its clause 66 engineer's decision procedure the final review by the employer's representative is final unless referred to conciliation or arbitration.

Reduction of extensions granted

Clause 44(5) states that the final review shall not result in any decrease in any extension granted under clause 44(3) or 44(4) – that is at the interim or due date stages. This inevitably raises the question, can the earlier assessments under clauses 44(3) and 44(4) produce reductions? Thus if an extension has been granted for an addition variation under clause 44(3) and there is subsequently a significant omission variation, is it possible to recognise this at a later clause 44(3) assessment or at a clause 44(4) assessment? The answer is probably – yes.

Sections

The provisions of clause 44 apply with equal effect to each and every section as well as to the whole of the works. Therefore each stage of procedure must be followed for each and every section.

Time limits

It is unlikely that failure by the employer's representative to act within the 14 day time limits prescribed in clauses 44(4) and 44(5) sets time at large. It is likely that they must be regarded as 'directory' only following the decision on a similar point in a building contract in the case of *Temloc Ltd* v. *Errill Properties Ltd* (1987).

14.7 Night and Sunday work

Clause 45 provides as in the ICE Sixth edition that none of the permanent works or the temporary works shall be carried out during the night or on Sundays without the permission of the employer's representative unless it is either:

- subject to any provisions in the contract to the contrary
- unavoidable

- necessary for saving life or property
- necessary for the safety of the works
- work which it is customary to carry out outside normal working hours or by shifts.

The clause does not apply to design work but only to 'permanent' work or 'temporary' work, neither of which by definition include design.

Provisions in the contract

In traditional contracting the provisions in the contract for night or Sunday working usually relate to pre-planned work with road or track closures and the like.

In the D and C Conditions the contractor has more scope than in traditional contracting to influence what is in the contract. Much of the detail will be in the contractor's submission.

Consequently unless the employer's requirements impose a veto on night or Sunday working the contractor can indicate in his submission the times he intends to work. These would, of course, be subject to acceptance of his submission in the first instance and compliance with statutory rules on noise and nuisance thereafter.

Work which is unavoidable

The wording of the clause is not absolutely clear with regard to work which is unavoidable. Two interpretations are possible. One is that unavoidable stands alone; the other is that unavoidable attaches to the words relating to the saving of life or property or for safety. If unavoidable does stand alone the question then becomes who decides what is, and what is not, unavoidable.

Steps to expedite progress

Note that under clause 46(2) if the employer's representative has served notice that the contractor shall take steps to expedite progress, permission to work at nights or on Sundays shall not be unreasonably refused.

14.8 Rate of progress

Clause 46(1) places a duty on the employer's representative to notify the contractor in writing when at any time, in his opinion, the rate of progress is too slow to ensure substantial completion by the due date. The contractor is then obliged to take such steps as are necessary – and to which the employer's representative may consent – so as to complete by the due date. The provisions apply to sections and to the whole of the works.

The clause commences with the words 'If for any reason which does not entitle the Contractor to an extension of time' and ends with the words 'The Contractor shall not be entitled to any additional payment for taking such steps'.

These provisions indicate that the employer's representative is obliged to constantly monitor progress and to press the contractor to complete on time. There is no sanction on the contractor if he ignores the notices issued other than in extreme cases of using the determination provisions of clause 65. In practice the contractor who is running late will make a commercial choice between incurring extra production costs or paying liquidated damages.

Constructive acceleration

Clause 46(1) contains a trap for the employer's representative which can be costly to the employer. It lies in the opening words of the clause which makes operation of the clause conditional upon the contractor not being entitled to an extension of time. If after giving notice to the contractor that he must take steps to expedite progress, the employer's representative or an arbitrator later grants an extension of time, the contractor may well be able to recover the costs of constructive acceleration.

The position, it is suggested, is even stronger for the contractor than for constructive acceleration resulting from failure by the employer's representative to operate properly the provisions of clause 44 in granting extensions of time.

Sir William Harris commenting in *New Civil Engineer* on the ICE Fifth edition in 1973 said:

> 'Not only may an employer's rights to liquidated damages be endangered if the extension of time provisions are not operated properly, but also he may find himself exposed to a claim for accelerated costs if the Contractor is pressed to complete by the

original contract date and eventually proves his entitlement to an extension.'

In both *Fernbrook Trading Co. Ltd* v. *Taggart* (1979) and *Perini Corporation* v. *Commonwealth of Australia* (1969) on civil engineering contracts, the judgments support the proposition that there is breach of contract if the engineer fails to grant extensions in reasonable time.

Permission to work nights or on Sundays

Clause 46(2) links back to clause 45 so that, if the employer's representative has served a clause 46 notice, he cannot then unreasonably refuse the contractor permission to work at night or on Sundays. The phrase 'on site' clarifies the need for permission.

Accelerated completion

Clause 46(3) has nothing to do with constructive acceleration discussed above. It relates only to 'requested' acceleration where the employer or the employer's representative requests the contractor to complete within a time less than that in the appendix or as extended.

The clause does no more than provide that, if the contractor agrees to accelerate, then any special terms and conditions of payment shall be agreed between the parties before action is taken. Clearly this cannot be binding on the parties if they choose to ignore it but it does have some practical effect in that the contractor can ask whatever price he likes.

Note that although the employer's representative can request acceleration, which in itself may seem odd, it is only the employer who can agree terms with the contractor.

Liquidated damages

Any accelerated completion date agreed under clause 46(3) may become binding for liquidated damages under the express wording of clause 47. This obviously needs to be considered in the financial agreement.

14.9 Liquidated damages for delay

The liquidated damages provisions of the D and C Conditions are the same as in the ICE Sixth edition and there is really nothing in design and construct contracting to require them to be different.

However, for the contractor there is more risk in design and construct contracting than in traditional contracting as to completion since delays and defaults in design do not qualify for extension of time. And there is some evidence in the building industry that where contractors offer their own completion times they regard such times as being of competitive significance and are inclined to understate the true time required.

Consequently, although liquidated damages provisions in design and construct contracts may be expressed the same as in traditional contracts, their usage may be greater.

General principles

The essence of liquidated damages is a genuine covenanted pre-estimate of loss. The characteristic of liquidated damages is that loss need not be proved.

Most standard forms of construction contract are drafted to permit the parties to fix in advance the damages payable for late completion. When these damages are a genuine pre-estimate of the loss likely to be suffered, or a lesser sum, they can rightly be termed liquidated damages. In short, liquidated damages are fixed in advance of the breach whereas general, or unliquidated damages, are assessed after the breach.

Reasons for use

There are sound commercial reasons for using liquidated damages. They bring certainty to the consequences of breach and they avoid the expense and dispute involved in proving loss.

Liquidated damages provisions are not solely for the benefit of the employer. They are also beneficial to contractors for they not only limit the contractor's liability for late completion to the sums stipulated, but they also indicate to the contractor at the time of his tender the extent of his risk. Thus, if a contractor believes that he cannot complete within the time allowed he can always build into his tender price his estimated liability for liquidated damages. All that the employer gets out of liquidated damages is relief from the burden of proving his loss and usually, in construction contracts, the right to deduct liquidated damages from sums due to the contractor.

Exhaustive remedy

It needs to be emphasised that if liquidated damages provisions are valid they provide an exhaustive and exclusive remedy for the specified breach. This was one of the points confirmed in the case of *Temloc* v. *Errill Properties Ltd* (1987) where it was said:

> 'I think it clear, both as a matter of construction and as one of common sense, that if ... the parties complete the relevant parts of the appendix ... then that constitutes an exhaustive agreement as to the damages which are ... payable by the contractor in the event of his failure to complete the works on time.'

The effect of this is that the employer cannot choose to ignore the liquidated damages provisions and sue for general damages. Nor can he recover any other damages for late completion beyond those specified.

If the provisions fail

It is well settled that when a liquidated damages clause fails to operate because it is successfully challenged as a penalty, or fails because of some defect in legal construction, act of prevention or other obstacle, then general damages can be sought as a substitute. Thus it was said in *Peak* v. *McKinney* (1970):

> 'If the employer is in any way responsible for the failure to achieve the completion date, he can recover no liquidated damages at all and is left to prove such general damages as he may have suffered.'

There is no firm ruling in English law that liquidated damages invariably act as a limit on any general damages which may be awarded in their place although this is generally thought to be the position.

Liquidated damages and penalties

There is a good deal of misunderstanding in the construction industry on the relationship between liquidated damages and penalties. Many believe that if the employer cannot prove his loss then any sum taken as liquidated damages must be a penalty. This is certainly not the case.

Providing the sum stipulated as liquidated damages is a genuine pre-estimate of loss it is immaterial whether or not the loss can be proved or even suffered; a point well illustrated by the case of *BFI Ltd* v. *DCB Integration Systems Ltd* (1987).

However if the stipulated sum is extravagant relative to any likely loss it may be held by the courts to be a penalty. English law does not allow the recovery of penalties and the courts will look at the stipulated sum irrespective of whether it is called liquidated damages or a penalty and if it is found to be a penalty will limit damages to the proven amount flowing from the breach. It is this which encourages contractors and their lawyers to find ways of challenging liquidated damages as penalties.

The rules on distinguishing penalties from liquidated damages laid down by Lord Dunedin in *Dunlop Pneumatic Tyre Co. Ltd* v. *New Garage and Motor Co. Ltd* (1915) are founded on the principle that liquidated damages must be a genuine pre-estimate of loss. Although an extravagant sum is the most obvious target for challenge, many cases have succeeded before the courts on technical arguments on whether or not the stipulated sum could, in all circumstances, be a genuine pre-estimate of loss.

Such arguments will have less chance of success in the future. The Privy Council held in the case of *Phillips Hong Kong Ltd* v. *The Attorney General of Hong Kong* (1993):

- the time when the clause should be judged was at the time of making of the contract not the time of the breach;
- so long as the sum payable was not extravagant having regard to the range of losses which could be reasonably anticipated, it can be a genuine pre-estimate of the loss;
- the argument that, in unlikely and hypothetical situations, the clause might provide damages greater than the loss actually suffered, was not valid;
- what the parties have agreed should normally be upheld.

The decision in the above case was followed in *J F Finnegan Ltd* v. *Community Housing Association Ltd* (1993) where it was held that the use of a formula to calculate the amount of liquidated damages was justified.

Clause 47(1) – liquidated damages for the whole of the works

Clause 47(1)(a) provides for liquidated damages where the whole of the works is not divided into sections. It says there 'shall' be included in the appendix a sum which represents the employer's genuine pre-estimate of damage likely to be suffered if the whole of the works is not substantially completed within the time allowed or as extended.

The reference to clause 46(3) indicates that the terms agreed for acceleration may include advancing the date on which damages become due.

Clause 47(1)(a) and the appendix both allow for liquidated damages to be expressed per week or per day.

It is usually better to use 'per day' since damages 'per week' cannot be proportioned down for a part week. Thus the contractor is only liable for each full week of delay if weeks are used. The not uncommon alteration of 'week' to 'week or any part thereof' is dangerous since that wording is open to challenge as a penalty.

It is not necessary for liquidated damages to be expressed in the same time periods as the time for completion.

Clause 47(1)(b) states the contractor's liability to pay liquidated damages if he fails to complete on time. Clause 47(5) details the methods of payment.

Completion of parts

The final paragraph of clause 47(1) is a provision for proportioning down the specified figure for liquidated damages when parts of the works have been completed prior to the whole.

The general position is that unless the contract provides a mechanism for the issue of partial completion certificates and a corresponding mechanism for proportioning down liquidated damages, the contractor remains liable for full damages up to the date of total completion.

However, where a contract does make provision for the issue of partial completion certificates, albeit on a consent basis, then it is essential to have corresponding provisions for proportioning down liquidated damages. The courts will not imply such a term if it is missing and the liquidated damages clause will then fall for uncertainty or as a penalty.

Clause 47(2) – liquidated damages for sections

Clause 47(2) repeats the provisions of clause 47(1) so that they apply when the whole of the works is divided into sections.

What happens is that the provisions in clause 47(1) for the whole of the works as a unity are effectively omitted and each section is given its own time and liquidated damages; with the 'remainder of the works' regarded as a further section.

Clause 47(2)(c) confirms a point fundamental to the scheme – namely that liquidated damages in respect of two or more sections may run concurrently.

Clause 47(3) – damages not a penalty

Clause 47(3) states that sums paid by the contractor pursuant to clause 47 are paid as liquidated damages and not as a penalty. This provision is of no legal effect.

The terminology itself is not decisive and if the courts find, as a matter of construction, that liquidated damages are penalties or vice versa they will award accordingly. Thus, Lord Dunedin in *Dunlop Pneumatic Tyre Co.* v. *New Garage and Motor Co. Ltd* (1915) said:

> 'Though the parties to a contract who use the words "penalty" or "liquidated damages" may prima facie be supposed to mean what they say, yet the expression is not conclusive.
>
> The court must find out whether the payment stipulated is in truth a penalty or liquidated damages.'

Clause 47(4) – limitation of damages

Clause 47(4)(a) allows for limitation of liability for sums payable as liquidated damages. This is, of course, at the discretion of the employer although it is not inconceivable that limits might be entered after negotiations with the contractor.

The statement in clause 47(4)(a) that if no limit is stated then 'liquidated damages without limit shall apply', means only that the contractor's liability is not stopped at a ceiling for prolonged delay.

Clause 47(4)(b) provides that if no sums are stated for liquidated damages in the appendix or if 'Nil' is written, then damages are not payable.

This means that the liquidated damages provisions do not become inoperative because of a blank or a nil entry; they remain in force but nothing can be recovered. The employer cannot sue for general damages for late completion because he has exhausted his remedy for the contractor's breach. This was one of the matters confirmed in the case of *Temloc Ltd* v. *Errill Properties Ltd* (1987) where the entry in the appendix to a JCT 80 contract was stated as £nil liquidated damages. The contract was finished late and the employer/developer, who was himself liable to the property purchaser for damages, sought to recover them as general damages from the contractor.

Clause 47(4)(b) goes further than the decision in the *Temloc* case by equating a blank entry with a nil entry. The effect of this is that if the employer genuinely wants the right to general damages as opposed to liquidated damages for late completion the whole of clause 47 needs to be deleted from the contract with corresponding amendments made to other clauses.

Clause 47(5) – recovery of damages

Clause 47(5) has two broad provisions:

- the employer may 'deduct' or require the contractor 'to pay' the amount due, and
- interest must be paid on sums reimbursed to the contractor for any overpayment.

It is worth noting that clause 47(5)(a) entitling the employer to deduct or require payment comes into operation without any condition precedent other than that in clause 47(1)(b) 'if the Contractor fails to complete ... within the time'.

Note also the wording 'The Employer may' at the start of clause 47(5). This indicates that the employer has a general option on whether to claim payment at all.

The final part of clause 47(5) provides that should extensions of time be granted which reduce the contractor's liability for damages, then any amounts overpaid shall be returned with interest. Not all standard forms allow for this.

The rate of interest is to be that provided for in clause 60(7) – 2% above base rate.

Clause 47(6) – intervention of variations

Clause 47(6) is an unusual provision, new to the ICE Sixth edition, but now overtaken by events. It is intended to deal with the question of what extension, if any, is due when delay for which the contractor is not responsible occurs after the due completion date – that is when the contractor is still progressing in culpable delay.

When the ICE Sixth edition and the D and C Conditions were produced there was no legal certainty as to whether the answer was:

- a gross extension to the date the delay ends
- a net extension for the period of delay
- time at large.

Clause 47(6) was introduced to avoid the argument by suspending damages for the period of delay.

However the decision in the recent case of *Balfour Beatty Building Ltd* v. *Chestermount Properties Ltd* (1993) makes the clause unnecessary. The court held that the contractor is entitled only to a net extension.

In the light of this decision and some untidy wording in clause 47(6) which will inevitably lead to difficulties the clause is perhaps best omitted.

One problem with the wording is that the grounds for suspending liquidated damages are too wide and go well beyond the events which would entitle the contractor to extension of time. Of particular concern is the phrase 'beyond the Contractor's control' which could bring in all time lost due to weather and the possibility of sub-contractor or suppliers delays.

Another problem is the phrase 'part of the Works' which suggests apportionment although the clause gives no guidance on how apportionment should be made.

Chapter 15

Completion and defects

15.1 *Introduction*

The provisions for completion and defects in the D and C Conditions are not significantly different from those in the ICE Sixth edition but greater emphasis is placed on the contractor's duty to supply operation and maintenance manuals.

The D and C Conditions have not followed the current trend in construction contracts of referring to 'taking-over' as opposed to 'completion' – although taking over is perhaps a better description of what actually happens. It indicates the true nature of the event.

At taking-over, or completion, responsibility for care of the works passes from the contractor to the employer – a matter of greater fundamental significance than the other consequences of completion – the limitation of liability for liquidated damages and the release of retention monies.

Low performance

Nor have the D and C Conditions taken on board any of the provisions to be found in some design and construct contracts for payment of damages or payment of a reduced price in the event of the completed works not reaching the specified standard.

Such provisions are, of course, essential for process and plant contracts but they also have a potential place in any contract intended to cover work which can be defined by performance parameters. It may be said that the D and C Conditions are intended only for conventional civil engineering works and that provisions for low performance would be superfluous. But that is to admit a measure of inflexibility in the application of the Conditions which competing forms of contract are seeking to avoid.

15.2 Definitions of completion

Various phrases are used to define completion:

- completion
- practical completion
- substantial completion.

The D and C Conditions use 'substantial completion' – a phrase returned to use in the ICE Sixth edition after inappropriate use of the phrase 'completion' in the ICE Fifth edition.

Completion

In its precise legal sense 'completion' means strict fulfilment of obligations under the contract. And when used in the context of 'entire contracts' which attract the doctrine of substantial performance failure to complete produces the apparently harsh result that no payment is due.

Thus in the famous case of *Cutter* v. *Powell* (1795) when the second mate on a ship bound for Liverpool from Jamaica died before the ship reached Liverpool his widow was unsuccessful in a claim for a proportion of his lump sum wages of 30 guineas.

But that is not to say that all lump sum contracts fall into the category of 'entire contracts'. Indeed if they did the risks for contractors would be immense since an employer unwilling to pay anything would only have to point to a modest default or item of unfinished work to escape the obligation of making payment.

Fortunately the courts take a practical view of building and construction contracts as illustrated by this extract from the judgment of Lord Justice Denning in the case of *Hoenig* v. *Isaacs* (1952) which concerned the decorating and fitting-out of a one room flat:

> 'In determining this issue the first question is whether, on the true construction of the contract, entire performance was a condition precedent to payment. It was a lump sum contract, but that does not mean that entire performance was a condition precedent to payment. When a contract provides for a specific sum to be paid on completion of specified work, the courts leap against a construction of the contract which would deprive the contractor of any payment at all simply because there are some

> defects or omissions. The promise to complete the work is, therefore, construed as a term of contract, but not as a condition. It is not every breach of that term which absolves the employer from his promise to pay the price, but only a breach which goes to the root of the contract, such as abandonment of the work when it is only half done. Unless the breach does go to the root of the matter, the employer cannot resist payment of the price. He must pay it and bring a cross-claim for the defects and omissions, or alternatively, set them up in diminution of the price. The measure is the amount which the work is worth less by reason of the defects and omissions, and is usually calculated by the cost of making them good.'

Nevertheless lump sum design and construct contracts come far closer to entire contracts than the traditional remeasurement contracts of civil engineering. And although it is not suggested that the D and C Conditions in their ordinary use could be construed as entire contracts it is certainly not inconceivable that in some circumstances the contractor's right to payment could be lost by failure to meet the employer's requirements. Such a situation is provided for in MF/1 which is principally for the supply and erection of plant and it can perhaps be implied into other fitness for purpose contracts.

Practical completion

Practical completion is the phrase commonly used in building contracts to define the point at which the works are fit to be taken over by the employer.

It is also used in the ICE Minor Works Conditions where it is stated:

> 'Practical completion of the whole of the Works shall occur when the Works reach a state when notwithstanding any defect or outstanding items therein they are taken or are fit to be taken into use or possession by the Employer.'

In the case of *Emson Eastern Ltd* v. *EME Developments Ltd* (1991) the court had to decide whether the issue of a certificate of practical completion under a JCT 80 contract constituted 'completion of the works' as mentioned in the determination clause of the contract. Judge Newey QC held that it did. He said:

'In my opinion there is no room for "completion" as distinct from "practical completion". Because a building can seldom if ever be built precisely as required by drawings and specification, the contract realistically refers to "practical completion", and not "completion" but they mean the same.'

From the cases of *H W Nevill (Sunblest) Ltd* v. *William Press & Sons Ltd* (1981) and *City of Westminster* v. *J Jarvis & Sons Ltd* (1970) the following rules to determine practical completion have been developed:

- Practical completion means the completion of all the construction work to be done.
- The contract administrator may have discretion to certify practical completion where there are minor items of work to complete on a *de minimis* basis.
- A certificate of practical completion cannot be issued if there are patent defects.
- The works can be practically complete notwithstanding latent defects.

Substantial completion

The phrase 'substantial completion' implies more flexibility than 'practical completion' and certainly the provisions in the D and C Conditions and the ICE Sixth edition indicate that it is not the *de minimis* principle which applies to outstanding work but whatever is acceptable to the employer's representative/engineer or is compatible with use or occupation of the works.

This flexibility has long been the cause of dispute under traditional ICE contracts with contractors giving notice of substantial completion and engineers requiring more work to be done before the issue of certificates of substantial completion.

There is nothing in the D and C Conditions to suggest that such disputes will be avoided and, if anything, they may increase. This is because the supervision arrangements for design and construct contracts place a heavier burden on the contractor to ensure compliance with standards and the completion of details. And the employer's representative will frequently not wish to be involved in monitoring and supervising work after the issue of a certificate of substantial completion. He may not have the resources to do so and he may not be paid to do so.

The probability is that substantial completion under the D and C Conditions will be adjudged more on the tests detailed above for practical completion than has been commonplace for traditional ICE contracts.

15.3 Clause 48 – certificate of substantial completion

The procedures in clause 48 for the issue of certificates of substantial completion are substantially the same as those in clause 48 of the ICE Sixth edition.

The only three changes to note are:

- employer's representative replaces engineer and engineer's representative
- tests prescribed 'by Statute' are added into clause 48(1)
- operation and maintenance instructions are to be provided to the employer prior to the issue of any certificate of substantial completion.

Clause 48(1) – contractor to give notice

Clause 48(1) provides that the contractor may give notice to the employer's representative when he considers that:

- the whole of the works, or
- any section with its own time in the appendix

has been:

- substantially completed, and
- has passed any final test prescribed in the contract or by statute.

Such notice is to be accompanied by an undertaking to finish any outstanding work in accordance with the provisions of clause 49(1).

As with the ICE Sixth edition, the responsibility for giving notice of completion falls on the contractor. And likewise it is expressed in the permissive term of 'may give notice' and not the mandatory term of 'shall give notice'.

The contractor's action under clause 48(1) is not strictly an application for a certificate of substantial completion, nor a request. Its purpose is to alert the employer's representative that in the opinion of the contractor, a certificate is due. The employer's representative must then take the appropriate action under clause 48(2).

Note that under clause 48(4) the contractor does 'request', and is entitled to, a certificate of substantial completion in respect of any substantial part of the works prematurely occupied or used by the employer.

Substantial completion of sections

Clause 48(1)(b) is quite specific in referring to any section in respect of which 'a separate time for completion is provided in the Appendix to the Form of Tender'.

Clearly, clause 48(1)(b) does not apply to parts of the works which, although identified in the contract, are not properly designated as sections. For further comment on this see below under 'premature use'.

Clause 48(2) – response of employer's representative

Clause 48(2) requires the employer's representative to respond within 21 days of the 'date of delivery' of the contractor's notice by either:

- issuing a certificate of substantial completion, or
- giving instructions in writing to the contractor specifying the work to be done before the issue of the certificate.

Date of certificate

The certificate, which is to be issued to the contractor and copied to the employer, is to state the date on which the works were, in the opinion of the employer's representative, substantially completed in accordance with the contract.

The certificate will, therefore, carry two dates – the date of issue and the date of substantial completion. The latter will normally be a prior date but note this does not upset clause 20 on care of the works which operates from the date of issue.

There is nothing to say that the date of substantial completion as adjudged by the employer's representative should be the same as that notified by the contractor. In fact the employer's representative can take the full 21 days from receipt of the contractor's notice not as administrative convenience, for which perhaps it is principally intended, but as time to allow the works to proceed to a more advanced stage of completion.

This can, and frequently does, lead to disputes on the deduction of liquidated damages and it then becomes a matter for the disputes resolution procedures of the contract.

Instructions on work to be done

Another frequent source of dispute over completion is disagreement on the amount of work necessary to permit the issue of a certificate of substantial completion.

The provision in clause 48(2)(b) requiring the employer's representative to give instructions in writing specifying the work which in 'his opinion' requires to be done suggests that subject to the test of reasonableness the opinion of the employer's representative prevails. However the test which applies in clause 48(2)(a) of 'substantially completed in accordance with the Contract' must be highly relevant to what is a reasonable opinion in clause 48(2)(b). So if it comes to a formal dispute before a conciliator or an arbitrator the issue is likely to be at what date were the works substantially completed in accordance with the contract.

There are, of course, some disputes where it is not a matter of the contractor getting on with the instructed works under protest and arguing later about the proper date for completion but more a matter of whether or not the instructed works can or should be undertaken at all. In such cases the issue then at conciliation or arbitration is usually of a technical nature.

As a general rule, the employer's representative needs to exercise caution in the use of clause 48(2)(b). It is certainly not, as its wording might suggest, a supplementary provision for issuing variation orders. So any instructions given under this clause should relate strictly to the contractual scope of works and not to late extras. And although there is not the problem in the D and C Conditions that can arise under the ICE Sixth edition, of instructions under clause 48(2)(b) producing claims under clause 13, nevertheless the very act of giving 'instructions' at a late stage of construction is potentially hazardous.

It would be better in the D and C Conditions for clause 48(2)(b) to simply require the employer's representative to specify the work to be done rather than require the giving of instructions.

Completion of work to be done

The final sentence of clause 48(2) states that the contractor is entitled to receive a certificate of substantial completion within 21 days of completion of the specified work to the satisfaction of the employer's representative.

There is no express requirement for the contractor to give notice of completion or make application for the certificate but it is obviously in the contractor's interest that he should do one or the other.

The reference to 21 days in the final sentence of clause 48(2) presumably has the same application as the reference to 21 days in the opening sentence of the clause. That is, the employer's representative is required to issue a certificate within 21 days stating the date of completion. Taken on its own the final sentence can be read as potentially adding 21 days to the date of completion.

Failure by the employer's representative to respond

Clause 48 is silent on the position which applies if the employer's representative fails to respond to the contractor's notice, in time or at all.

The contractor has a strong case for arguing that once the 21 days has passed it is too late for the employer's representative to give instructions on works to be completed; and he has an equally strong case for arguing that his notice should be effective in fixing the date of substantial completion.

Operation and maintenance instructions

Clause 48(3) provides that prior to the issue of any certificate of substantial completion under clause 48(2) the contractor is to provide the employer with operating and maintenance instructions in sufficient detail for the works being taken over to be:

- operated
- maintained

- dismantled
- reassembled
- adjusted.

Although there is no matching provision in clause 48 of the ICE Sixth edition a similar provision is to be found in clause 7(6)(b) of those Conditions in relation to any part of the permanent works designed by the contractor.

However, it is an interesting point that whereas the provision in the ICE Sixth edition says 'No certificate under Clause 48 ... shall be issued', the provision in the D and C Conditions is specifically related to any certificate issued under clause 48(2).

From this it would seem to follow that certificates issued under clause 48(4) – premature use or occupation, or certificates issued under clause 48(5) – substantial completion of other parts, are not caught by the requirements of clause 48(3).

In the case of premature use or occupation this is understandable since it would not be possible to impose an obligation on the contractor in respect of such an event. In the case of substantial completion of other parts, the issue of a certificate is very much a discretionary power of the employer's representative and no doubt one factor relevant to the exercise of that discretion will be the availability of operation and maintenance manuals or instructions.

Clause 48(4) – premature use by the employer

Clause 48(4) provides for the issue of a certificate of substantial completion for any 'substantial' part of the works which has been occupied or used by the employer 'other than as provided in the Contract'.

The procedure is simply that the contractor shall request and the employer's representative shall issue. There is no reference to 'the opinion of the employer's representative'; there is no need for the contractor to give an undertaking on outstanding work; nor is there any provision for the employer's representative to refuse until further work has been completed. And, as mentioned above, there is no requirement on the provision of operating and maintenance manuals.

Although this clause is identical in wording to that in the ICE Sixth edition its impact is likely to be different. It is one thing for an employer to take premature occupation or make premature use of works which have been designed by his own engineer; it is quite another to move prematurely into works designed by the contractor. Firstly, there is the question of consent to consider; then there is the question of overall design integrity.

There can be little doubt that in design and construct contracting premature occupation or use should be avoided if possible.

Premature use

As to exactly what constitutes premature use, the wording of clause 48(4) is potentially misleading. The clause says 'any substantial part' and 'other than as provided in the Contract'. Arguments about what is a 'substantial part' are likely to be arguments on facts; but arguments on what is meant by 'provided in the Contract' are more likely to be of a legal nature.

The Conditions themselves only impose completion obligations on the contractor in respect of the whole of the works or sections with specified times in the appendix. The contract may elsewhere in the employer's requirements state obligations in respect of the completion of parts; but it must be doubted if this is what is meant by 'provided in the Contract' in clause 48(4). This is because, as mentioned above, Clause 48(2) deals only with the whole of the works or sections. It does not deal with parts. Therefore, if the employer does prematurely occupy or use a part which is not a section, the certificate of substantial completion will be issued under clause 48(4) whether or not that part is provided for in the contract.

Clause 48(5) – completion of other parts

Clause 48(5) confers on the employer's representative a discretionary power to issue a certificate of substantial completion for any part of the works which has:

- been substantially completed
- passed any final test prescribed in the contract.

Upon the issue of such a certificate the contractor is deemed to have undertaken to complete any outstanding work during the defects correction period.

This is, in many respects, an unusual provision, because it does not depend upon any of the criteria expressed elsewhere in clause

48 for the issue of a certificate of completion. There is no mention of:

- notice or application from the contractor
- an undertaking to complete outstanding work (except that this is deemed to be given)
- any obligation to complete outstanding work other than during the defects correction period
- provision of operating and maintenance instructions
- occupation or use by the employer
- the part being a substantial part.

The effect is that the employer's representative has the widest possible discretion. But it is a discretion to be used prudently and only, one would expect, after consultation with the employer – for one effect of the certificate is to pass responsibility for care of that part of the works to the employer under clause 20(1)(b).

Reinstatement of ground

Clause 48(6) provides that a certificate of substantial completion for a section or a part shall not be deemed to certify completion of any ground or surfaces requiring reinstatement unless the certificate expressly so states.

This is a practical provision having regard to the contractor's continued occupation of the site.

15.4 Clause 49 – outstanding work and defects

Clause 49 sets out, in similar detail to the ICE Sixth edition, the contractor's obligations for the completion of outstanding work and defects.

Clause 49(1) – outstanding work

The time allowed to the contractor to complete outstanding work is either:

- that set by agreement with the employer's representative under clause 48(1), or

- if no time is so agreed, as soon as practicable during the defects correction period.

The provision for agreement of a time in which to complete outstanding work was new to the ICE Sixth edition and clearly in practice it will have some influence in both the ICE Sixth edition and the D and C Conditions in decisions on whether or not to allow contractors their certificates of substantial completion. It has long been a point of discontent with engineers and employers that contractors, having been given certificates of completion, leave the site and put off finishing outstanding works until the very end of the defects correction period.

Contractors may argue that the new provision amounts to unfair pressure – either agree to complete all outstanding work within a set time or no certificate will be issued until it is complete. Employers are more likely to say that is how things should be. But in any event, the procedure for agreement applies only to clause 48(1) and not to certificates issued under clauses 48(3) and 48(4).

A point which is none too clear is what remedy, if any, does the employer have if the contractor defaults on his agreement to complete outstanding work in a set time?

It is possible that clause 49(4) may address the issue allowing the employer to engage others to complete the work but, as shown below, there is a strong case for saying that clause 49(4) applies only to defects. If that is the case the employer has no obvious contractual remedy but he could sue for damages for breach if the defects impeded his use or occupation of the works. That apart the contractor's agreement may be no better than his word.

Clause 49(2) – work of repair

The first part of clause 49(2) requires the contractor to 'deliver up to the Employer' the works, or any section or part for which a certificate has been issued, as soon as practicable after the end of the relevant defects correction period, in the condition:

- required by the contract (fair wear and tear excepted), and
- to the satisfaction of the employer's representative.

The second part of clause 49(2) continues:

'To this end the Contractor shall repair amend reconstruct rectify and make good defects of whatever nature notified to him in writing by the Employer's Representative during the relevant Defects Correction Period or within 14 days after its expiry as a result of an inspection made by or on behalf of the Employer's Representative prior to its expiry.'

As with the ICE Sixth edition, which has slightly different wording, it is far from clear precisely what the contractor's obligation is in respect of time. The first part of the clause says 'as soon as practicable after the end of the relevant Defects Correction Period'. The second part of the clause can be taken as an obligation to complete work of repair either during the relevant period or within 14 days of its expiry.

However, a better interpretation of the second part of the clause is that the employer's representative 'may' notify defects during the defects correction period but 'must' notify them, in any event, within 14 days of its expiry. And the inspection made by the employer's representative must be made before expiry of the defects correction period.

On this basis the contractor's obligation remains no more than to undertake work of repair as soon as practicable after the end of the defects correction period.

Inspection by the employer's representative

As to the situation when the employer's representative fails to give notice of defects within 14 days after expiry of the defects correction period, or fails to make his inspection prior to its expiry, contractors may argue that they are then relieved of any obligation to deal with defects. It is unlikely that this argument is wholly effective since the obligation in the first part of clause 49(2), and the contractor's general obligation as expressed in clause 8 and elsewhere to complete the works in accordance with the contract may take precedence. Nevertheless this is not a matter which any employer's representative should risk putting to the test.

'Defects of whatever nature'

The phrase 'defects of whatever nature' in clause 49(2) indicates that the clause is not confined to repairs and defects for which the contractor is responsible. This is supported by clause 51(1) which expressly states that variations may be ordered at any time up to the end of the defects correction period 'for the whole of the Works'.

Clause 49(3) – cost of repairs

Clause 49(3) confirms the obvious point that works of repair are to be carried out at the contractor's cost if they are due to his failure to comply with any of his obligations under the contract, but otherwise they are to be paid for as if they are additional work.

The words due to 'use of materials or workmanship not in accordance with the contract or to neglect' found in the corresponding clause in the ICE Sixth edition are omitted from the D and C Conditions. The change is not thought to be significant since the words appear to be superfluous even in the ICE Sixth edition.

Where the contractor is entitled to payment, clause 49(3) provides only that the work should be valued as 'additional work'. This leaves open the possibility that it can be valued either as a variation or on a daywork basis.

Clause 49(4) – contractor's failure to carry out repairs

Clause 49(4) entitles the employer, in the event of the contractor's failure to do 'any such work as aforesaid', to carry out such work himself. The employer can recover the cost from the contractor if the work should have been carried out at the contractor's own expense.

It is not clear whether the phrase 'any such work as aforesaid' covers both outstanding works under clause 49(1) and remedial works under clause 49(2) or whether it applies only to the latter.

There is an argument for saying that because the phrase 'such work' appears in clause 49(3) relating to repairs, then the phrase only applies to repairs in clause 49(4). However that would leave the employer without a practical remedy in the event of uncompleted work.

The difficulty with including outstanding work within the scope of clause 49(4) is that it entitles the employer to act within the defects correction period once the period described as 'as soon as practicable' in clause 49(1) is adjudged to have elapsed. This may open up much scope for dispute but it may, nevertheless, be the correct interpretation.

However, before taking any action under clause 49(4) with the intention of charging the contractor, the employer needs to consider carefully how he is going to justify his action if a formal dispute arises. By acting, the employer will probably destroy the best evidence for his case and he will need to rely on records of the defects he has put right.

15.5 Clause 50 – contractor to search

Clause 50 obliges the contractor to carry out:

- searches
- tests, or
- trials

as may be necessary to determine the cause of any:

- defect
- imperfection, or
- fault

if so required, in writing, by the employer's representative.

The clause follows the wording of clause 50 in the ICE Sixth edition except that the phrase 'under the directions of the Engineer' is omitted. The thought presumably is that under the D and C Conditions it is up to the contractor to decide how best to carry out his searches, tests and trials.

There is nothing expressly in clause 50 to restrict its application to the defects correction period. However the reference in its last line to 'expense in accordance with Clause 49' implies that it is intended for use in the defects correction period. It can also be said that clause 36(3) (checks and tests) and clause 38(2) (uncovering and making openings) deal with tests during the construction period.

But against this there is the point that clauses 36, 38 and 50 serve slightly different purposes:

- clause 36(3) deals with routine testing for quality
- clause 38(2) deals with uncovering work put out of view
- clause 50 deals with searches in respect of patent defects.

So clause 50 is undoubtedly a useful provision. The question is whether or not it is necessary for the construction period. Perhaps the position in the D and C Conditions is that the contractor has an implied obligation to determine the cause of patent defects which emerge during the construction period and there is no need for an express provision such as clause 50 to impose the obligations.

Costs of searching

Clause 50 provides that the contractor is to bear the costs of searching etc if the defects are found to be his liability; otherwise the employer bears the costs.

Under the D and C Conditions with the contractor's extensive design and construct responsibilities there would not seem to be much scope for the contractor to avoid liability. The employer would, no doubt, certainly believe this to be the case; particularly as the contractor is obliged to accept responsibility for the employer's design under clause 8(2)(b).

However, there is, of course, the matter of unforeseen conditions and the acceptance of responsibility by the employer under clause 12. Other matters could be use or occupation by the employer.

Works of repair

Where the contractor is liable, clause 50 obliges the contractor to repair and rectify at his own expense 'in accordance with clause 49'. This latter phrase presumably involves timescale.

Clause 50 says nothing on the ordering or execution of repairs where the contractor is not liable but the employer's representative has powers under clauses 51 and 62 which may be adequate.

15.6 Clause 61 – defects correction certificate

Clause 61(1) provides for the employer's representative to issue to the employer (with a copy to the contractor) a single 'Defects Correction Certificate' when:

- all outstanding work referred to under clause 48 is completed
- all work of repair, amendment and making good is completed
- all other faults referred to under clauses 49 and 50 have been completed.

The defects correction certificate is to state the date on which the contractor has completed his obligations 'to construct and complete' the works to the satisfaction of the employer's representative.

The omission of any reference to the contractor's 'design' obligation in clause 61(1) may reflect the practical nature of the defects correction certificate or it may simply be due to a direct transfer of wording from the ICE Sixth edition.

Clause 61(1) makes it absolutely clear that where there is more than one defects correction period, as there will be of course with sections or with partial handovers, then it is only on the expiry of the last period that the defects correction certificate becomes due.

Unfulfilled obligations

Clause 61(2) states that the issue of the defects correction certificate shall not be taken as relieving either of the parties of their liabilities to one another arising out of the contract.

This gives the certificate much lower status than similar certificates in other construction contracts where such certificates finalise certain matters so they cannot thereafter be disputed. All that the clause 61 certificate does is:

- establish the last date for the submission of the contractor's final account under clause 60(4), and
- terminate the defects correction period thus ending any remaining rights of the contractor to enter the site to remedy defects.

Manuals and drawings

Clause 61(3) has no equivalent in the ICE Sixth edition. It requires the contractor to submit manuals and drawings prior to the issue of the defects correction certificate.

Under clause 61(3)(a) the contractor is to submit to the employer's representative for approval:

- one set of draft operation and maintenance manuals, and
- associated records and drawings.

The detail is to be sufficient to enable the employer to:

- operate
- maintain
- dismantle
- reassemble
- adjust

the permanent works.

Under clause 61(3)(b), once the employer's representative has given approval, the contractor is to supply three sets of the manuals and drawings to the employer.

15.7 Clause 62 – urgent repairs

Clause 62 deals with remedial work or other work or repairs which, in the opinion of the employer's representative, are urgently necessary.

The clause refers specifically to accident or failure but it is comprehensive in its scope by its reference to 'or other event'. And it expressly says 'either during the carrying out of the Works or during the Defects Correction Period'.

The clause provides that, firstly, the employer's representative shall inform the contractor with confirmation in writing. Thereafter, if the contractor is unwilling or unable to carry out the work or repair the employer can make other arrangements to undertake the work.

The cost of the work so done is charged to the contractor if it is his liability, otherwise the cost falls on the employer.

Chapter 16

Variations

16.1 Introduction

The apparent inability of construction projects to proceed without change has long been a source of cost to employers and profit to contractors. It might be said that an employer has no one to blame but himself for this and the solution is straightforward. Do not on any account change your mind.

But in traditional contracting that is a gross simplification of the position. The majority of variations are not so much the result of the employer's change of mind but are the result of his designer's change of details. One only has to look at the coverage of what constitutes a variation in clause 51 of the ICE Sixth edition to realise that almost any communication between the engineer and the contractor is potentially a variation. And the situation is little different under other forms of contract.

In the much quoted building case of *Simplex Concrete Piles Ltd* v. *St. Pancras Borough Council* (1958) an architect who assented to a contractor's proposals for remedial works was held to have issued a variation. More recently under the ICE Fifth edition in the case of *English Industrial Estates* v. *Kier International Construction Ltd* (1991) it was held that an engineer's letter instructing the contractor to crush all suitable hard arisings deprived him of the option of importing suitable fill and was therefore a variation.

And under traditional ICE conditions even silence or refusal by the engineer to issue a variation is no guarantee of the employer's protection. Thus in *Yorkshire Water Authority* v. *Sir Alfred McAlpine & Son (Northern) Ltd* (1985) it was held that incorporation of a method statement into a contract made it a specified method of construction and the contractor was entitled to a variation for an alternative method of working. In *Holland Dredging (UK) Ltd* v. *Dredging & Construction Co Ltd* (1987) it was held that the contractor was entitled to a variation order for measures in rectifying a shortfall in backfill to a sea outfall pipe arising from loss of dredged material.

Variations in design and construct

One of the principal incentives towards design and construct contracting is that it leaves the employer far less exposed to claims for variations. Providing the employer's requirements do not stray too far into the realm of detail it is the contractor who takes broad responsibility for detail and any necessary changes thereto.

Changes of mind, of course, remain as variations. And since they probably have to be valued against a lump sum contract price and may impact on the contractor's design in ways never intended, or expected, such variations are likely to be expensive. In the building industry, employers are often advised – stick with your original proposals and make any changes after the contractor has completed the works – it will cost less.

This may not be a practical solution in civil engineering projects but it does illustrate that lump sum design and construct contracting is not suitable for projects where fundamental decisions on the employer's requirements have still to be made, or significant change of mind variations are probable.

16.2 Variations generally

A broad definition of a variation is work which is not expressly or implicitly included in the original contract price.

In the absence of an express provision in the contract giving the employer the power to order variations the contractor is not obliged to undertake them. The contractor's general obligation is merely to complete the work specified in the contract and such work as can reasonably be inferred for completion.

Reasons for variation

Variation clauses are included in contracts principally to permit the employer to make changes and to order extras but they may have other purposes:

- to authorise the contract administrator to order variations
- to entitle the contractor to a variation if the specified work proves impossible to construct
- to define the scheme for the valuation of variations
- to prescribe procedures to be followed to recover payment

- to preclude payment for unauthorised variations
- to set financial limits on authorised variations.

Not infrequently variation clauses also seek to define what is permitted as a variation and to indicate the scope of the contractual provisions.

Scope of variation clauses

By and large contractors welcome variations as a means of increasing revenue so disputes on whether particular variations are outside the scope of the contract are not too common. When contractors do complain it is usually in the hope of either a better valuation of the varied work than the contract permits or of getting the entire contract price set aside and payment made on a quantum meruit basis. See the observations in Chapter 17 on this and the case of *McAlpine Humberoak* v. *McDermott International* (1992).

The limits of the variation provisions can usually be deduced from the nature of the contract work and the wording of the variation clause. Thus in the ICE Sixth edition a variation is either to be necessary for completion of the works or desirable for completion and improved functioning. Clearly the varied works need to have some relationship to the original work. But it does not follow that they will be a variation simply because they are carried out at the same time as the original works and on the same site. In a case under the ICE Fifth edition, *Blue Circle Industries plc* v. *Holland Dredging Co. (UK) Ltd* (1987) it was held that work in attempting to form an island bird sanctuary with dredged material was not within the scope of the variation clause.

Omission variations

Variation clauses in construction contracts usually allow for omissions as well as additions. But that does not amount to freedom for the employer to take work out of the contract and to give it to others or to omit work simply to save money. See the case of *Carr* v. *Berriman Pty Ltd* (1953).

The general rule is that the contractor is entitled to carry out the work in the contract. To be genuine variations, omissions must fulfil the requirements of the variation clause in the same way as additions.

16.3 *Variations in design and construct contracts*

In traditional contracting where the contractor basically constructs what he is told to construct, variations flow naturally from the employer, or his representative, to the contractor. The contractor may occasionally put forward his own ideas of how the works can be improved, but such ideas, if accepted, will usually be formalised as variations from the employer.

In design and construct contracting many of the decisions on how the works should be built come from the contractor and the employer's scope for making variations is correspondingly limited.

Terminology

Many standard design and construct forms of contract recognise that traditional variations have been displaced by the transfer of design responsibility and they use the word 'changes' instead of 'variations'. Such is the case in JCT 81 and The Department of Transport Design and Construct form.

The intention, perhaps, is to emphasise that the contract works should be in accordance with either the employer's requirements, the contractor's submission, or a combination of both, and it is only by changes to these that the works can be varied.

The D and C Conditions retain the word 'variations' but this probably results from the parentage of the form and the familiarity of civil engineers with the concept of 'variations'.

Limitations

It is one thing for an employer to issue variations on a contract for which he carries responsibility for design but quite another to impose changes on a contractor who carries the design burden. Not only is there the question of the effect of the changes on the integrity of the design but there is also the question of the contractor's capability to process the changes.

Some standard design and construct forms such as the IChemE Red and Green books and MF/1 place financial percentages limits, 25% and 15% in these cases, on the value of variations which can be imposed by the employer without the contractor's consent. It is also commonplace to require the contractor to give notice to the employer if he considers that compliance with any variation would prejudice fulfilment of any of his obligations under the contract.

The D and C Conditions do not directly address any of these issues but the phrase in clause 51(1) 'the Employer's Representative shall have power after consultation with the Contractor's Representative to vary the Employer's Requirements' may effectively deal with them all.

Valuations

Design and construct contracts have three principal methods of valuing variations:

- by contractor's quotation
- by reference to the contract sum analysis/schedule of rates
- by contractor's quotation based on the contract sum analysis/schedule of rates.

Many standard forms seek to include in the quotation all the consequential costs of the variations and the effects on time by reference to any extension considered necessary. The quotation, if accepted, then becomes binding on the contractor and he takes the risk on both cost and time. Naturally, in such circumstances, the contractor is entitled to build into his quotation allowances for the risks carried.

In the D and C Conditions the valuation of variations is intended to be quotation based, with both value and delay agreed before work is commenced. The rules however are not as precise as in some other contracts.

Abortive design costs

One effect of imposing limits on the value of variations or seeking agreement in advance of commencement on the price, is that the employer's proposed changes may be no more than speculative until the price is known. The contractor in the meantime may have incurred substantial design costs.

There is no uniform approach in design and construct contracts on whether the contractor should be reimbursed such costs. The IChemE forms allow the contractor to recover his costs plus an allowance for profit for preparing quotations whether or not the variations proceed.

The D and C Conditions are silent on the matter.

Disputes

In most traditional contracts the contractor is obliged to comply with instructions from the engineer (or architect) and if he is aggrieved he can seek redress through the formal disputes procedure in the course of time. In design and construct contracting, disputes on whether or not to proceed with a variation may have more immediate effect.

To cope with this some design and construct forms have introduced 'expert' or 'adjudicator' dispute resolution procedures. This is not the case in the D and C Conditions.

Contractor's changes

Even in traditional contracting not all proposals for change come from the employer. The contractor has ideas as well. At its extreme in design and construct contracting the employer may have only the broadest notion of what he requires – all the detail comes from the contractor in his submission or proposals. It follows that suggestions for changes are more likely to come from the contractor as he develops his scheme than the employer.

Accordingly it is usual in design and construct contracts for the variations provisions to accommodate both employer's changes and contractor's changes.

The D and C Conditions refer only to variations in the employer's requirements. The contractor's submission is fixed by definition at the award of the contract. Further comment on this is made in Chapter 3.

16.4 Variations in the Design and Construct Conditions

Power to vary

Clause 51(1) gives the employer's representative power to vary the employer's requirements – but only after consultation with the contractor's representative. The clause does not place an express duty on the employer's representative to order variations necessary for completion of the works – as does the ICE Sixth edition and earlier versions of the traditional form. To have included such a duty would no doubt have been seen to undermine the principle that the contractor is responsible for design, construction and completion.

Nevertheless, what may at first sight appear as a major change from the ICE Sixth edition may be something of an illusion. The contractor's obligation under clause 8(1) is to design, construct and complete the works 'save in so far as it is legally or physically impossible'. If the contractor can show that the employer's requirements contain such impossibility, the employer's representative will have to vary the requirements if an alternative solution exists. Otherwise the employer would face abandonment of the works. This is not far removed from the position in the ICE Sixth edition where the engineer is obliged to vary the works to overcome impossibility in his own design or specified methods.

The contractor under the D and C Conditions has, of course, to deal himself with impossibility in his own design and this may entail changes outside the scope of clause 51. But again see the comment in Chapter 3 on this.

Consultation

The ICE Sixth edition requires the engineer to 'consult' the contractor only in the valuation of variations. In the D and C Conditions the power to make variations is itself linked to consultation.

The *Concise Oxford Dictionary* defines the act of consulting as: deliberation; conference. To consult is to have deliberations with; seek information or advice from; take into consideration. These definitions indicate two possible meanings for 'consultation' as used in clause 51(1). Firstly, there is the practical point. The employer's representative should seek information or advice from the contractor (about the impact of the proposed change on the contractor's design, for example) before ordering a variation. Secondly, there is the fairness point. The employer's representative should not order a variation without taking into consideration the effect on, and the views of, the contractor.

The question of whether one or both applies is likely to be of relevance only to a dispute about variations ordered without consultation. The contractor may well be entitled to refuse to comply with a variation on practical or safety grounds if it has been made without his involvement. His case for refusal on the grounds that the variation was not fair would be less strong.

Variations to the employer's requirements

Both clause 51 (the ordering of variations) and clause 52(2) (the valuation of ordered variations) deal only with variations to the employer's requirements. Unlike the variation clauses of most other design and construct forms there is no reference to changes in the contractor's submission.

The logic of this is probably that the contractor's obligation is to design, construct and complete the works to the employer's requirements. Therefore any changes in the contractor's obligation can only be expressed as changes in the employer's requirements. That is to say, clause 51 seems to be concerned only with changes in the contractor's obligations.

Clearly the contractor can put forward proposals for change which if adopted by the employer can be formalised as variations; the contractor then has changed his obligations. What is less clear in the D and C Conditions is whether the contractor can change the details in his submission where the generality of the employer's requirements is such that details are not obligations.

It certainly appears that if such changes are permitted in the Conditions they are outside the scope of clause 51.

Variations by consent

The possibility that the Conditions contemplate variations additional to those to the employer's requirements under clause 51 can be deduced from the wording of clause 36(6) 'whenever a variation is ordered or consented to by the Employer's Representative' and by clause 2(4)(c)(ii) 'any variation of or in the Works'.

Note also clause 6(2)(c) by which the contractor has to notify the employer's representative if he wishes to modify any design or drawing to which consent has been given.

Moreover, whereas clause 52(1) on the valuation of variations refers to 'the work as varied', the scheme in clause 52(2) applies specifically to 'variations ordered . . . in accordance with Clause 51'.

Perhaps the intention is that variations in the contractor's details can be made with consent but any financial implications must be settled by agreement on the basis of the contractor's quotation. This is effectively what is to be found in other forms of contract which directly address the matter.

Scope of clause 51

Clause 51(1) states that 'variations may include additions and/or omissions'.

This, it will be noted, is much shorter than the coverage detailed in clause 51(1) of the ICE Sixth edition which includes changes in quality, character, position, levels, timing etc. Again the explanation is that variations in the ICE Sixth edition are likely to be changes in detail whereas in the D and C Conditions, details if they exist as contractual obligations, are likely to be within the employers' requirements and covered by the broad scope of 'change'.

Consequently, although the D and C Conditions express the coverage of variations quite briefly the extent of the coverage can be just as wide as in the ICE Sixth edition if the employer's requirements are themselves expressed in detail.

Variations in the defects correction period

The question of whether variations can be ordered after the issue of the certificate of substantial completion is put to rest in these Conditions, and in the ICE Sixth edition, by the express provision in clause 51(1) permitting variations to be ordered 'up to the end of the Defects Correction Period for the whole of the Works'.

The provision is to be welcomed in principle but it can be criticised for apparently permitting variations to be ordered for a part or section of the works after the defects correction period for that part or section has expired.

Variations to be in writing

Clause 52(2) states the familiar provision that all variations shall be ordered in writing; but it then allows for oral instructions by reference to clause 2(5).

Under clause 2(5)(a) the contractor is required to comply with oral instructions. Under clause 2(5)(b) if the contractor confirms in writing an oral instruction and it is not contradicted in writing forthwith it is deemed to be an instruction in writing.

The question arising from this is – can the contractor claim a variation if, neither he nor the employer's representative confirmed the oral instruction in writing? The answer is probably yes. Clause 2(5)(a) places a duty on the employer's representative to confirm whereas the contractor has a right but not a duty to confirm.

Variations not to affect the contract

Clause 51(3) provides, as in the ICE Sixth edition, that no variation shall vitiate or invalidate the contract.

This seems to be no more than a precaution against legal challenge to the validity of the contract. Certain contracts outside construction do state the opposite – that they are invalidated by variations. And, of course, even in construction contracts the underlying premise of many global claims, or claims for quantum meruit, is that the contract has been invalidated by the scale of variations.

Fair and reasonable value

Clause 51(3) also provides that a fair and reasonable value of all variations shall be taken into account in ascertaining the contract price.

The phrase 'fair and reasonable' is an interesting addition to that in the ICE Sixth edition wording. The ICE Sixth edition provides a well defined code for the valuation of variations based principally on bill rates. Contractors often argue, with some justification, that it is unfair to value variations on rates which may in themselves be inadequate. That however, whether fair or unfair, is the contract they have entered into.

In the D and C Conditions, although the contractor is entitled to claim a fair and reasonable valuation, that will not protect him from the consequences of an inadequately priced quotation for the variation. Fair and reasonable means fair and reasonable to both parties and the contractor cannot commit the employer to a variation on one price and then demand a higher price at a later date.

Contractor's default

The final provision of clause 51(3) is that the contract price is not to be adjusted for variations necessitated by the contractor's default.

This might seem so obvious that it need not be said. But as shown by the *Simplex Concrete Piles* case mentioned at the start of this chapter the courts are inclined to take contracts to mean what they say. Thus a bare statement that all variations are to be valued in the contract price may be taken to mean just that, whatever the cause.

16.5 Valuation of variations

Ordered variations

The D and C Conditions provide the following rules for the valuation of ordered variations:

- all valuations to be fair and reasonable – clauses 51(3) and 52(2)(d);
- whenever possible valuations to be agreed before any variation is ordered or work commences – clause 52(1);
- valuations not so agreed to be based on the contractor's quotation or negotiations thereon – clauses 52(2)(a) and 52(2)(b);
- valuations to include for delay consequences and costs – clauses 52(1) and 52(2)(a);
- failing agreement on valuation the employer's representative may make a fair and reasonable assessment; although only possibly for the purposes of interim valuations – clause 52(2)(d).

Consent variations

It is not clear if these rules also apply to consent variations – that is those variations requested by the contractor for changes in detail which do not involve changes in the employer's requirements. Clause 6(2)(c) clearly contemplates such changes and requires the contractor to obtain consent before making modifications but it does not address the financial implications.

On one line of thought, that which says the contractor's submission imposes binding obligations, application of the rules would seem to be both sensible and just.

But on the alternative line of thought, that which says the contractor's submission is always secondary to the employer's requirements – a proposition supported by clause 5(1)(b) – there should be no valuation of consent variations. The argument is that the contractor has priced to comply with the employer's requirements and unless those requirements change there should be no change in the contract price.

The effect of this if it is correct, which on balance is probable, is that the contractor takes the full cost benefit of savings he can identify within his submission. This may surprise employers who think they have been offered and have accepted a higher specification or superior scheme than their requirements demand. The solution is to deal with such matters prior to the award of the contract and not afterwards.

Omission variations

The valuation of ordered omission variations raises similar difficult issues.

Clause 51(1) expressly states that variations to the employer's requirements may include omissions. Clause 51(3) says that the fair and reasonable valuation of all variations shall be taken into account in ascertaining the contract price. It would seem to follow that an ordered omission variation should result, in most circumstances, in a reduction in price.

The problem lies in clause 52 which, in detailing the rules for the variation of ordered variations, refers only to cost and delay.

It is possible to argue that if the contractor fails to offer a reduction in price for an omission the employer's representative can make a fair and reasonable valuation under clause 52(2)(d). This would be in line with the thinking of the drafting committee which was that omission variations would normally lead to a reduction in price.

But again there is a contrary argument to the effect that in a lump sum contract the price can only be adjusted by express provisions and in the absence of such provisions, omissions do not lead to a reduction.

This is an important matter and it is unfortunate that the Conditions have left scope for uncertainty. Users of the Conditions may well wish to clarify their intentions perhaps by adding into clause 52 a simple statement that the valuation of all omission variations shall be on a fair and reasonable basis.

16.6 Valuation procedure in clause 52

Clause 52(1)

Clause 52(1) deals with the ideal situation where the value and delay consequences of a variation are agreed before the order is issued or work starts. Note, however, that this clause is activated by a request from the employer's representative for a quotation.

It is suggested that the phrase 'value and delay consequences' is intended to cover both direct and consequential effects of the variations thereby going some way towards eliminating general delay and disruption claims on a cumulative basis. The wording of the clause, if this is the intention, is not as firm as it might be. Some contracts emphasise the point for the avoidance of doubt that generalised claims based on variations are not permitted.

Clause 52(2)

The opening phrase of clause 52(2) 'In all other cases' is slightly ambiguous. It may refer to cases where there has been no request for a quotation under clause 52(1); or no agreement under that clause; or to both. It needs to refer to both since otherwise failure to agree under clause 52(1) is left without any procedure for resolution or payment. However if it does apply to failure to agree the effect may be that the contractor is entitled to submit another quotation after receipt of the variation order and to discard his initial quotation.

Clause 52(2)(a) – cost

Under clause 52(2)(a) the contractor is required to submit his quotation 'As soon as possible'. However, it is worth noting that the additional payment provisions of clause 53 although expressly not applicable to clause 52(1) apply to clause 52(2). Those provisions require notice to be given within 28 days. This, then, may be the limit of what is 'As soon as possible'.

The phrase in clause 52(2)(a)(i) 'having due regard to any rates or prices included in the Contract' will have little effect in a lump sum contract unless there is a contract price analysis and/or a schedule of rates.

Clause 52(2)(a) – delay

The intention of clause 52(2)(a)(ii) requiring the contractor to give his estimate of delay would seem to be to fix the extension of time due for the variation. Thus in clause 52(2)(b) the estimate is to be accepted or negotiated upon. But acceptance may not give any extension agreed the same finality as an agreement on cost.

This is because the employer's representative has a duty to review extensions of time under clause 44 at the due date for completion and again on substantial completion and he may have an obligation to deal with any inadequacy in the estimate. Added to this there is the point that clause 52(2)(c) requires only the contract price to be amended upon agreement with the contractor. It says nothing as to time.

Clause 52(2)(c)

This clause confirms the finality of an agreement as to the value of a variation. The contractor, of course, needs to allow in his price for the risk this entails.

Clause 52(2)(d)

In the absence of agreement on price the employer's representative is obliged to notify the contractor of what in his opinion is a fair and reasonable valuation.

The wording of clause 52(2)(d) is somewhat obscure and it may be that this valuation is for the purposes of interim payment only. Settlement of the valuation beyond this stage would seem to be by agreement of the parties or under the conciliation or arbitration procedures of clause 66.

Clause 52(3) – daywork

Clause 52(3) permits the employer's representative to order in writing that any:

- additional work, or
- substituted work

which in his opinion is:

- necessary, or
- desirable

shall be executed on a daywork basis.

This clause simply provides the power to order dayworks. The rules for payment are given in clause 56(3).

Chapter 17

Claims for additional payments

17.1 *Introduction*

For employers one of the major attractions of design and construct procurement is that it offers the prospect of greater certainty of price than traditional contracting. This has been the principal driving force behind the growth of design and build contracting in the building industry and it is likely to be the main stimulus towards design and construct contracting in civil engineering.

The twin pillars supporting the proposition of greater certainty of price are single point responsibility and lump sum pricing.

Single point responsibility

There is no doubt that many of the most common causes of contractor's claims are eliminated when the employer is not liable to the contractor for the deficiencies of the designer of the works. Claims for late supply of information, errors in drawings and variation of working details become matters to be settled directly between the contractor and his designer.

Under traditional contracting the employer may have a remedy to recover his outgoings on such claims from the designer as breach of contract – usually on the basis of negligent performance. Anecdotal evidence suggests that the practice has become widespread in recent years, with professional indemnity insurers picking up the bill. For obvious reasons the subject is not discussed too openly by professional design firms; but changing approaches to public accountability and commercial awareness in the last decade or so have certainly hardened the attitudes of many employers. For some it is now a matter of policy that no one is let off the hook.

But whatever the extent of this form of recovery under traditional contracting it may well be overshadowed compared with the scale of recovery exacted from designers under design and construct contracting. Professional design firms will find contractors hard taskmasters on standards of performance. And any deficiencies which would have resulted in contractors' claims against the employer under traditional contracting are likely to be vigorously pursued by contractors against their own externally appointed designers. The result is likely to be a steep rise in claims on professional indemnity policies. If that in turn raises premiums the cost eventually will fall back on employers but at least it will do so in a predictable manner and certainty of price will be maintained.

Lump sum pricing

In traditional ICE conditions of contract the employer does not warrant that the quantities are correct and he takes the financial risk on the quantities as measured. Such contracts are known as measure and value contracts. As far as certainty of price is concerned they obviously suffer from a major defect.

In lump sum pricing the contractor takes the risk that his assessment of the quantities involved in the works is accurate. If he is wrong he stands the cost. The employer gets the benefit of the firm price.

The D and C Conditions are not drafted specifically for lump sum pricing but that is how they will mainly be used. If they are used with quantities on a remeasurement basis a good deal of thought needs to be given to who prepares the quantities and who takes the risk on their accuracy.

Grounds for additional payment

Although the D and C Conditions when used with lump sum pricing should offer the employer a fair measure of certainty of price – providing he can desist from ordering variations – there are nevertheless 31 clauses (43 sub-clauses) in the Conditions which provide for, or otherwise contemplate, adjustment of the lump sum. This is only five less than in the ICE Sixth edition. The various clauses are scheduled later in this chapter.

Many of the clauses are by way of indemnities or relate to contingency matters which the employer should have allowed for in his budget. But clause 12 on unforeseen physical conditions remains as in the ICE Sixth edition and the employer's liability for the performance of his representative is much the same as the employer's liability for the engineer under the ICE Sixth edition with the notable exceptions of late supply of drawings and incorrect setting out data.

Settlement of claims

A point of significant difference between the D and C Conditions and the ICE Sixth edition is the very limited role of the employer's representative in the settlement of claims.

The scheme under the ICE Sixth edition is broadly as follows. Under a specific provision the contractor becomes entitled to such cost as may be reasonable. The contractor gives notice under clause 52(4) and applies for payment under clause 60. The engineer certifies under clause 60 the amount he considers is due. Either party can dispute the amount so certified and request an engineer's decision under clause 66. That decision is final and binding unless referred to conciliation or arbitration within a set time. Thus the engineer has the power and the duty to settle contractual claims within the disputes procedure.

In the D and C Conditions there is no provision for a decision under clause 66 and there is nothing in the Conditions to give finality to amounts certified by the employer's representative. The effect of this is that the employer's representative is not empowered to settle claims for additional payment. Any dispute must be settled by agreement between the parties, by conciliation or by arbitration.

For further comment on this see Chapter 20 on the resolution of disputes.

The impact of claims

Employers will rightly be attracted to design and construct contracting by the reduced scope for claims and the greater certainty of price. But a word of caution needs to be added. When things go wrong on a design and construct contract they can go wrong in a very big way. The employer no longer has control through his engineer of the design process and if he has to change his requirements to accommodate the unforeseen or the unexpected the cost in time and money is likely to be greater than in traditional contracting.

The ability of the contractor to escape from his lump sum price to cost-plus reimbursement is discussed in greater detail below but the point is recognised in process engineering and certain other industries that if the employer's requirements are so uncertain that he is likely to end up paying cost-plus he might as well contract on that basis in the first place. Consequently the IChemE forms of contract, much used in the water industry, have a lump sum version (the Red book) and a cost reimbursable version (the Green book).

The D and C Conditions could be modified to provide for cost reimbursement but the necessary amendments would be more than superficial and would certainly be best left to lawyers.

17.2 Legal basis of claims

Claims for additional payments fall into one of two categories – claims under the terms of the contract or claims at common law. They are sometimes referred to as contractual claims and extra-contractual claims.

Contractual claims

The legal basis of claims under the terms of a contract is the contract itself.

The parties to a contract are free, subject to considerations of legality, to agree whatever terms they wish and whatever arrangements they choose for payments and additional payments. They are not bound to observe the common law rules on risk, remoteness or measure of damages. And in setting down express terms the parties are illustrating their intention that some or all of the following considerations should apply to the various or separate provisions of their contract:

- common law rights confirmed
- common law rights amended
- common law rights excluded
- additional rights given
- notice requirements specified
- administrative rules stated
- interim payments permitted
- cost, loss or expense defined

- powers to value and to settle given
- valuation rules outlined
- finality of decisions indicated
- limitation on actions accepted
- settlement of disputes procedures stated
- damages to be liquidated.

With such an extensive range of variations to common law rules available it is hardly surprising that standard forms of contract offer a marked difference of balance between employers and contractors. Amongst design and construct forms used by civil engineers at one end of the spectrum there is the IChemE Red book with its limitations on contractor's liability and at the other end there is The Department of Transport form with its emphasis on contractor's liability.

The D and C Conditions fall in between the two extremes with the balance following broadly that set by the ICE Sixth edition and its forerunner the ICE Fifth edition which was generally held to be fair to both parties. The contractual provisions in themselves, although lacking some of the precision of the corresponding ICE Sixth edition provisions, will be neither unfamiliar nor unexpected to experienced ICE Fifth and ICE Sixth edition users. A greater source of difficulty is likely to be adapting from measure and value to lump sum principles.

Common law claims

At common law there are four clearly defined ways of recovering money:

- for breach of contract
- on the basis of quantum meruit
- under implied contracts
- for the tort of negligence.

Additionally it is sometimes possible to recover on an *ex gratia* promise.

Apart from scope of entitlement the most obvious difference between common law claims and contractual claims is that the contract itself will rarely contain any provisions for submission, valuation or payment of common law claims. Thus under the ICE Sixth edition the power of the engineer is generally limited to dealing with contractual claims and only under clause 66 is the engineer required, or permitted, to deal with wider matters. In practice, however, the engineer does frequently acquire an involvement in the settlement of extra-contractual claims. But he usually does this not as engineer to the contract but in the role of agent of the employer. In this capacity he is not required to be impartial or to make a fair valuation; he is seeking the best settlement for the employer.

Under the D and C Conditions the employer's representative fulfils many of the familiar duties of the engineer but he has no authority or requirement to make a decision under clause 66 as a condition precedent to arbitration. He is required under clause 60(4) to issue a certificate stating the amount which in his opinion is finally due under the contract but clearly this does not include valuation of extra-contractual claims. Again, in practice the employer's representative is likely to become involved in such claims; firstly because he will often be the negotiating agent of the employer; and secondly because contractors do not always distinguish between contractual claims and extra-contractual claims in their submissions or, following the lead of lawyers, they claim for both in the alternative.

It probably matters less in the D and C Conditions if the role of the employer's representative as contract administrator becomes blurred with his role as employer's agent. There is no finality attached to his valuations and no express requirement that he should be impartial.

Alternative claims

In most contracts, and the D and C Conditions are no exception, there is a measure of overlap between contractual rights of claim and common law rights. Clause 42(3) on failure to give possession of the site is an obvious example. The question then arises – are the contractual rights the limit of the contractor's entitlement or do common law rights stand alongside as an alternative? The point is relevant in relation to giving, or failing to give, notice of intention to claim; and in relation to the amount recoverable where the contractual amount is less than that recoverable as common law damages.

The position varies from form to form but the answer to the question may be found in the contract itself. Thus in JCT 80 the contract says with reference to claims:

> 'The provisions of this Condition are without prejudice to any other rights and remedies which the contractor may possess.'

However, in contrast the IMechE/IEE standard form, MF/1, states:

> 'The Purchaser and the Contractor intend that their respective rights, obligations and liabilities as provided for in the Conditions shall be exhaustive of the rights, obligations and liabilities of each of them to the other arising out of, under or in connection with the Contract or the Works, whether such rights, obligations and liabilities arise in respect or in consequence of a breach of contract or of statutory duty or a tortious or negligent act or omission which gives rise to a remedy at common law.'

In D and C Conditions, as in the ICE Sixth edition, the position is not clear. There is no express inclusion of common law rights and no express exclusion. Consequently there are two schools of thought on the matter. One argues that common law rights can only be excluded by express terms; the other argues that specific provisions on a particular matter represent that exclusive remedy.

Whatever the answer it is important to note that the point is only relevant to claims where there is both a contractual provision for claim and a corresponding common law right of claim. It has no relevance to claims such as clause 12 for unforeseen physical conditions which are additional to breach of contract.

Breach of contract

Claims for breach of contract can rely on an express term or an implied term.

In construction contracts the more important obligations of the parties are usually set out as express terms. Thus the contractor shall design, construct and complete to a defined quality within a set time; the employer shall give possession of the site, supply information and pay on certificates. But not all the obligations can be fully expressed and the law will imply such terms as are necessary for the business efficacy of the contract.

Many implied terms derive from the principle of prevention which states that one party cannot impose a contractual obligation on the other party where he has impeded the other in the performance of that obligation. Lord Justice Vaughan Williams in the case of *Barque Quilpue Ltd* v. *Bryant* (1904) put it this way:

> 'There is an implied contract by each party that he will not do anything to prevent the other party from performing a contract or to delay him in performing it. I agree that generally such a term is by law imported into every contract.'

Where there is breach of an express term the starting point for a claim is straightforward. But a claim which relies on an implied term has first to overcome the hurdle of proving the existence of the implied term. Thus in the *Glenlion* case mentioned in Chapter 7 the contractor alleged that there was an implied term that the employer would provide information in time to meet the requirements of a shortened programme of works. In that case the contractor failed.

Quantum meruit

Claims for quantum meruit – meaning 'what it is worth' are sometimes called quasi-contractual claims. They are highly popular with contractors wishing to escape from the rigidity of a lump sum or from contract rates towards payment on a cost-plus basis.

Strictly speaking the phrase 'quantum meruit' applies to the law of restitution for the value of services rendered where there is no contractual entitlement to payment. But it is also commonly used to describe claims made under a contract for a fair valuation or a reasonable sum. The division between the two is not always clearcut as this comment by Mr Justice Goff in *British Steel Corporation* v. *Cleveland Bridge & Engineering* (1981) shows:

> 'a quantum meruit claim – straddles the boundaries of what we now call contract and restitution; so the mere framing of a claim as a quantum meruit claim, or a claim for a reasonable sum, does not assist in classifying the claim as contractual or quasi-contractual.'

In that case British Steel had supplied steel nodes to Cleveland Bridge for a job in Saudi Arabia but a formal contract was never concluded because neither party would agree to the other's terms of business. It was held that British Steel were entitled to be paid a reasonable sum for the work done.

In the broad sense, claims for quantum meruit can be made where:

- no contract is ever concluded
- the contract is unenforceable
- the contract is void for mistake
- the contract is discharged by frustration
- the contract is for a lump sum and the employer prevents completion
- no price is fixed in the contract
- the work undertaken falls outside the scope of the contract
- the contract provides for payment of a reasonable sum or a fair valuation.

However, it must be said that the law of restitution is complex and the cases where claims succeed frequently turn on particular facts. Such cases often prove to be of little lasting authority.

Thus the old case of *Bush* v. *Whitehaven Port and Town Trustees* (1888) much quoted by contractors in support of winter working claims was held by the House of Lords in *Davis* v. *Fareham UDC* (1956) not to be authority for any principle of law. Bush, the contractor, was prevented by late possession of the site from commencing work on a pipeline contract until winter when wages were higher and the work was more difficult to construct. The court held that as a summer contract had been contemplated Bush was entitled to recover his extra expenditure as damages or to be paid for all the work on quantum meruit.

The proposition that the *Bush* case was said to support was that where the circumstances of a contract are altered the contract price is no longer binding and can be replaced by quantum meruit. This has obvious appeal to contractors. But two recent cases show that the courts are not disposed to award quantum meruit where there are contractual provisions for compensation.

In *Morrison-Knudsen* v. *British Columbia Hydro and Power Authority* (1985) the contractor could have terminated the contract because of serious breaches by the employer. Instead the contractor elected to continue working and claimed for the value of work on quantum meruit. The court held that such an award could not be given as an adequate remedy was available under the contract.

In *McAlpine Humberoak v. McDermott International* (1992) the contract for the deck structure of an off-shore drilling rig was let on a lump sum basis. The contractor claimed extra costs arising from variations. The judge at first instance relying, rather surprisingly, on the *Bush* decision, held that the contract had been frustrated by the extent of the variations and awarded the contractor quantum meruit. The Court of Appeal overturned this decision and held:

- the variations did not transform the contract or distort its substance or identity
- the contract provided for variations
- the contractual machinery for valuing variations had not been displaced
- an award of quantum meruit could not be supported.

Implied contracts

Claims under implied contracts are comparatively rare for the reasons discussed in section 10.2 of Chapter 10 under collateral contracts. Nevertheless with the scope for recovering economic loss under the tort of negligence now severely restricted there will inevitably be increased attempts to prove contractual relationships where no formal contract exists.

The barrier to such claims is usually the absence of consideration – a necessary requirement of English common law contracts. However the courts do appear to be adopting a policy of finding consideration wherever possible.

In *Blackpool and Fylde Aero Club* v. *Blackpool Borough Council* (1990) a tender for the concession to operate pleasure flights from an airport was mistakenly believed to have been delivered late and was not considered by the Council. The Aero Club sued for negligence and breach of contract. The Court of Appeal held that the facts of the case justified the discovery of an implied contract – that in return for the tenderer expending the time and expense of preparing a tender the Council would open and consider the tender. The Council were held liable for damages.

In *Lester Williams* v. *Roffey Brothers and Nicholls (Contractors) Ltd* (1989), Williams, a sub-contractor, agreed to carry out the carpentry work in a block of flats for a lump sum of £20 000. Williams got into financial difficulties, partly because the price was too low, and Roffey, the contractor, promised to pay a further sum of £10 300 in instalments to avoid delay on the contract. Roffey paid £1 500 of this and then no more. Williams stopped work and sued for the balance. Roffey argued that there was no consideration for the extra payment as Williams was only doing what he was contractually obliged to do. The Court of Appeal held that the promise was contractually binding on Roffey. This was the explanation given by Lord Justice Glidewell:

> 'The present state of the law on this subject can be expressed in the following proposition:

(i) if A has entered into a contract with B to do work for, or to supply goods or services to, B in return for payment by B; and

(ii) at some stage before A has completely performed his obligations under the contract B has reason to doubt whether A will, or will be able to, complete his side of the bargain; and

(iii) B thereupon promises A an additional payment in return for A's promise to perform his contractual obligations on time; and

(iv) as a result of giving his promise, B obtains in practice a benefit, or obviates a disbenefit; and

(v) B's promise is not given as a result of economic duress or fraud on the part of A; then

(vi) the benefit to B is capable of being consideration for B's promise, so that the promise will be legally binding.'

The tort of negligence

Until quite recently it appeared to be possible, given appropriate circumstances, for a contractor to claim damages in tort as an alternative to breach of contract. Lord Denning MR had said in *Photo Productions Ltd* v. *Securicor Transport Ltd* (1980):

> 'If the facts disclose the self-same duty of care arising both in contract and in tort and a breach of that duty, then the plaintiff can sue in either contract or tort.'

However, Lord Scarman in *Tai Hing Cotton Mill* v. *Liu Chong Hing Bank* (1986) cast doubt on this saying:

> 'Their Lordships do not believe that there is anything to the advantage of the law's development in searching for a liability in tort where the parties are in a contractual relationship.'

And in *Surrey Heath Borough Council* v. *Lovell Construction* (1988) it was conceded that if a contract expressly deals with the subject of a claim there is no room for a parallel duty in tort.

There is therefore little scope for a contractor to claim against an employer in tort and recent decisions have virtually eliminated the possibility of successfully claiming against the contract administrator.

Two cases, *Arenson* v. *Arenson* (1977) and *Sallis* v. *Calil* (1988) had indicated that a contractor might be able to recover damages against a negligent certifier. But the Court of Appeal in *Pacific Associates* v. *Baxter* (1988) ruled that an engineer had no duty of care in respect of economic loss to the contractor where the arbitration clause in the contract provided the contractor's remedy.

Ex gratia payments

On literal translation ex gratia payments are those made 'out of kindness'. In practice they are regularly made to settle claims on a compromise basis or without prejudice to liability.

Additionally contractors occasionally lodge claims on an ex gratia basis where they are losing money and can find no proper grounds for claim. The employer has no obligation to pay and cannot be sued. However, as seen in *Williams* v. *Roffey* above, making a promise to pay and then defaulting can give the contractor a legal entitlement to recovery.

17.3 *The amount recoverable*

To establish a successful common law claim it is necessary to overcome certain legal obstacles:

- proving that loss or damage occurred
- showing that the loss was caused by the events relied on
- establishing the amount of the loss
- showing that the loss is not too remote
- showing that the loss could not have been mitigated by reasonable conduct.

The same obstacles apply in principle to contractual claims but the items on remoteness and mitigation usually apply with less effect since the basis of recovery is often defined with precision in the contract.

Burden of proof

The burden of proof rests with the claimant on the principle that he who asserts must prove.

Many construction claims fail because the contractor has difficulty on this elementary point. Not every breach by the employer results in loss to the contractor.

Causation

Linking loss to a particular cause is anything but straightforward. And in construction contracts where there are frequently concurrent or overlapping problems and delays the difficulties are often formidable.

It is beyond the scope of this book to examine the techniques of retrospective delay analysis (RDA) but broadly it can be said there are three main approaches:

- first in time
- dominant cause
- apportionment.

Experts in retrospective delay analysis develop these and other approaches into complex models with computers regarded as an essential tool.

Remoteness of damage

The law does not allow a claimant to succeed in every case where damage follows breach but draws a practical line by excluding that which is too remote.

The guiding principles of remoteness applying to cases of breach of contract derive from the judgment of Baron Alderson in the very old case of *Hadley* v. *Baxendale* (1854).

The rule in *Hadley* v. *Baxendale* is taken as having two branches and is commonly expressed as:

> 'Such losses as may fairly and reasonably be considered as either arising:
>
> (1st rule) "naturally", i.e. according to the usual course of things,
>
> or
>
> (2nd rule) "such as may reasonably be supposed to be in the contemplation of both parties at the time they made the contract, as the probable result of breach of it".'

The test of remoteness laid down by Baron Alderson was reformulated in the classic judgment of Lord Justice Asquith in the case of *Victoria Laundry (Windsor) Ltd* v. *Newman Industries Ltd* (1949) in terms of foreseeability.

However, in *Czarnikow* v. *Koufos* (1969) known as *The Heron II* the House of Lords moved away from the foreseeability test to one of assumed common knowledge. The effect of this on the law and a summary of the law as it now stands was admirably expressed by the Court of Appeal of New Zealand in *Bevan Investments* v. *Blackhall & Struthers* (1978) as follows:

> '(1) The aggrieved party is only entitled to recover such part of the loss actually resulting as may fairly and reasonably be considered as arising naturally, that is according to the usual course of things, from the breach of the contract.
> (2) The question is to be judged as at the time of the contract.
> (3) In order to make the contract-breaker liable it is not necessary that he should actually have asked himself what loss was liable to result from a breach of the kind which subsequently occurred. It suffices that if he had considered the question he would as a reasonable man have concluded that the loss in question was "liable to result".
> (4) The words "liable to result" should be read in the sense conveyed by the expressions "a serious possibility" and "a real danger".'

Measure of damages

The principles applied by the courts in measuring damages date back to the case of *Robinson* v. *Harman* (1848) where it was stated:

> 'The rule of common law is that where a party sustains a loss by reason of a breach of contract, he is, so far as money can do it, to be placed in the same situation, with respect to damages, as if the contract had been performed.'

This rule is, of course, subordinate to the rule on remoteness.

There are, then, two further points to consider:

- the rules on mitigation of loss by reasonable conduct and
- the rules on the recovery of wasted expenditure.

Wasted expenditure

Some judicial guidance on wasted expenditure can be gained from the case of *C & P Haulage* v. *Middleton* (1983). In that case, a motor repairer executed certain works to premises he occupied for his business to render them suitable for his purpose and sued for wasted expenditure when his lease was terminated in breach of contract.

The Court of Appeal held that he could not succeed as he had suffered no loss of profit because he had found alternative accommodation and the earlier 'wasted' expenditure would have been spent anyway even if the contract had not been broken.

The general point is that a claim for damages is not intended to improve one's position on what it would have been without any breach. In short, a claim is not a device for turning loss into profit.

Mitigation of loss

It is said that a claimant has a duty to mitigate his loss. This is true to the extent that the claimant seeks to recover his loss as damages, but it does not follow that an injured party in a breach of contract should have his conduct determined by the breach.

The following extracts from legal judgments explain this. Viscount Haldane in *British Westinghouse Electric & Manufacturing Co. Ltd* v. *Underground Electric Railways of London Ltd* (1912) said that:

> 'A plaintiff is under no duty to mitigate his loss, despite the habitual use by the lawyers of the phrase "duty to mitigate". He is completely free to act as he judges to be in his best interest. On the other hand, a defendant is not liable for all loss suffered by the plaintiff in consequence of his so acting. A defendant is only liable for such part of the plaintiff's loss as is properly to be regarded as caused by the defendant's breach of duty.'

Sir John Donaldson, Master of the Rolls, in *The Solholt* (1983) said:

> 'The fundamental basis is thus compensation for pecuniary loss naturally flowing from the breach; but this first principle is qualified by a second, which imposes on a plaintiff the duty of taking all reasonable steps to mitigate the loss consequent on the breach, and debars him from claiming any part of the damage which is due to his neglect to take such steps.'

There is clearly wide scope for debate on how far the concept of 'neglect to take such steps' should apply. It is doubtful, for example, that it extends to expenditure of further monies which might or might not be recoverable but it probably does include taking-up reasonable offers and applying practical steps.

17.4 *Global claims*

Strict application of the rules of causation and the burden of proving loss create serious difficulties in construction claims. Frequently the contractor finds it difficult to identify how much of his losses are his own responsibility and how much should be charged to the employer; and that is before he starts the task of trying to allocate his costs to individual claimable events.

Claiming on a global basis obviously alleviates these difficulties; but the question is how acceptable is it for the contractor to roll-up his costs and put them forward as the composite loss for a bundle of claims?

The case of *J Crosby & Sons Ltd* v. *Portland Urban District Council* (1967) is often quoted as authority for the practice but it needs to be treated with caution. It applies only to quantification and then only when the contractual machinery is no longer effective.

In the *Crosby* case an arbitrator set out his award for the court to consider as a special case. Mr Justice Donaldson, as he was then, not only approved the global award but also included in his judgment an extract from the arbitrator's findings.

> 'The result, in terms of delay and disorganisation, of each of the matters referred to above was a continuing one. As each matter occurred its consequences were added to the cumulative consequences of the matters which had preceded it. The delay and disorganisation which ultimately resulted was cumulative and attributable to the combined effect of all these matters. It is therefore impracticable, if not impossible, to assess the additional expense caused by delay and disorganisation due to any one of these matters in isolation from the other matters.'

On the strength of those words and far wider application of Mr Justice Donaldson's supportive judgment than stands examination a great deal of money has since changed hands; much of it on the dubious proposition that total cost less tender sum is the amount of the employer's liability.

It took until 1985 and the judgment of Mr Justice Vinelott in the case of *London Borough of Merton* v. *Stanley Hugh Leach Ltd* (1985) before the potentially fatal flaw in global claims was publically revealed; namely that the method of calculation often relieves the contractor of any burden of his own additional costs or pricing errors. But the *Merton* case far from containing the global claims approach opened it up to new horizons for the case with its wide publicity on so many other points of interest confirmed that the principles of the *Crosby* case applied to JCT contracts as well as to ICE contracts. And moreover the judge in describing a global award as a supplement to contractual machinery suggested that in appropriate circumstances the architect had a duty to ascertain global loss.

However, something of a retreat from those cases has taken place in recent years. Principally because employers have sought to apply the global approach in bringing actions against their designers.

Firstly came the Hong Kong case of *Wharf Properties Ltd* v. *Eric Cumine Associates* which came before the Privy Council in 1991 and which confirmed that the *Crosby* and the *Merton* cases applied only to quantification and that those cases gave no relief to a claimant from the obligation to plead his case with particularity. Or, as better expressed in the words of Lord Oliver of Aylmerton:

> 'Those cases establish no more than this, that in cases where the full extent of extra costs incurred through delay depend upon a complex interaction between the consequences of various events, so that it may be difficult to make an accurate apportionment of the total extra costs, it may be proper for an arbitrator to make individual financial awards in respect of claims which can conveniently be dealt with in isolation and a supplementary award in respect of the financial consequences of the remainder as a composite whole. This has, however, no bearing upon the obligation of a plaintiff to plead his case with such particularity as is sufficient to alert the opposite party to the case which is going to be made against him at the trial.'

In the *Wharf* case the employer had paid out 317 million or so Hong Kong dollars to the contractor in claims and he was trying to recover this from his architect whom he claimed had failed to properly manage, control, co-ordinate, supervise and administer the project. However the best the employer could do with the information available was to make a global approach to his pleadings. Neither the *Crosby* nor the *Merton* cases came to his assistance. Indeed Lord Oliver brought out an aspect of both cases which has been widely overlooked – that those cases say only what it might be proper for an arbitrator to do; not what it might be proper for a claimant to do.

Two other cases have followed the *Wharf* case with employers trying to recover monies paid out on cost over-runs; *Mid Glamorgan County Council* v. *Devonald Williams and Partner* (1991) and *Imperial Chemical Industries plc* v. *Bovis and Others* (1992).

Mr Recorder Tackaberry in the *Mid Glamorgan* case gave a useful summary of the legal position following the *Wharf* case:

> '1. A proper cause of action has to be pleaded.
> 2. Where specific events are relied on as giving rise to a claim for monies under the contract then any pre-conditions which are made applicable to such claims by the terms of the relevant contract will have to be satisfied, and satisfied in respect of each of the causative events relied upon.
> 3. When it comes to quantum, whether time based or not, and whether claimed under the contract or by way of damages, then a proper nexus should be pleaded which relates each event relied upon to the money claimed.
> 4. Where, however, a claim is made for extra costs incurred through delay as a result of various events whose consequences have a complex inter-reaction which renders specific relation between the event and time/money consequence impossible and impracticable, it is possible to maintain a composite claim.'

The *ICI* case arose out of the refurbishment of ICI's premises in Millbank where the cost rose from an original estimate of approximately £30M to over £53M. The contract was let on a management basis and ICI sued the management contractor, the architects and the consulting engineers in respect of £19M of cost over-runs and associated fees. A global claim was made against each of the defendants. As with the *Wharf* and the *Mid Glamorgan* cases much of the interest of the case is in the approach of the courts to inadequate pleadings but again, as in the *Merton* case the potential for absurdity in global claims was revealed.

The global claim for abortive work in respect of hundreds of items amounted to £840 211. It was said that apportionment was impossible. The defendants asked, if they had a complete defence to all the items save for two minor ones 'circuits need changing' and 'fire bell repositioned', what monetary consequences would flow from these two items? The reply was to the effect 'If any of the events is not proven at trial, the only consequence is that the actual sum paid will fall to be distributed between a lesser number of events, not that the total recoverable will be less'. His Honour Judge Fox-Andrews found it 'palpable nonsense that £840 000 could be the cost of repositioning a fire bell'.

17.5 Express provisions for additional payment

The various express provisions for additional payment to the contractor in the D and C Conditions are as follows:

Clause	**Description**	**Entitlement**
5(1)(c)(ii)	instructions from the employer's representative in explaining and adjusting ambiguities or discrepancies in the employer's requirements	amount of such cost as may be reasonable beyond that reasonably to have been foreseen plus profit on additional work
6(1)(b)	failure by the employer's representative to issue information required for the design or construction of the works within a reasonable period following notice	any reasonable cost which may arise
6(2)(d)	failure by the employer's representative to give consent (or notice of withholding consent) within a reasonable period following submission of design or drawings for consent	any reasonable cost which may arise
8(3)	failure by the employer's representative to approve (or indicate withholding of approval) within a reasonable period of submission or resubmission of the contractor's quality plan and procedures	any reasonable cost which may arise

Clause	Description	Entitlement
9	execution of the contract agreement	cost
12(6)	the contractor encounters physical conditions or artificial obstructions which could not reasonably have been foreseen by an experienced contractor	reasonable cost of carrying out additional work plus a reasonable percentage for profit and the reasonable cost of delay or disruption to the rest of the works
18(2)(a)	boreholes arising from a clause 12 situation	as clause 12
18(2)(b)	boreholes arising from a variation	as clause 52
18(3)	boreholes ordered in writing by the employer's representative	as a variation, contingency or prime cost as appropriate
20(3)(b)	loss or damage to the works arising from any of the excepted risks	expense
20(3)(c)	loss or damage to the works arising in part from an excepted risk	cost as apportioned
22(3)	damage to persons or property arising from exceptions which are the responsibility of the employer	indemnity against claims, costs, etc
22(4)	damage to persons or property arising in part from exceptions	indemnity for proportion of liability
26(2)	statutory fees in relation to the design and construction of the works and rates and taxes payable in respect of the site or other structures used temporarily or exclusively for the works	all sums properly payable and paid by the contractor

Clause	Description	Entitlement
26(3)(b)	instructions issued by the employer's representative to ensure compliance with Acts of Parliament, statutory regulations and bye-laws	a variation to be ordered
26(3)(c)	employer warrants that all necessary planning permission for the permanent works and specified temporary works has been obtained	payment implied
28(2)	infringement of patents etc for which the employer takes responsibility	indemnity against claims, costs etc
29(4)	noise, disturbance or pollution which is the unavoidable consequence of carrying out the works	indemnity against claims, costs etc
30(3)(c)	damage to bridges or highways caused by the transportation of materials which the contractor has taken all reasonable means to prevent	indemnity against claims, costs etc
31(2)	affording facilities for other contractors, employees, statutory bodies etc employed in carrying out work not in the contract	such cost as may be reasonable plus reasonable profit in respect of additional work
32	orders of the employer's representative in respect of fossils and antiquities	expense
36(5)	cost of samples not clearly intended by or provided for in the contract	cost

Clause	Description	Entitlement
36(7)	cost of tests not clearly intended by or provided for in the contract unless such tests reveal workmanship or materials not in accordance with the contract	cost
38(2)	uncovering work etc as directed by the employer's representative which is found to be in accordance with the contract	cost
40(1)	suspension of work on the written order of the employer's representative which is not either provided for in the contract; necessary by reason of weather; nor contractor's default; nor necessary for proper construction or safety	extra cost plus profit in respect of additional work
40(2)	suspension lasting more than three months not either provided for in the contract; nor by reason of default of the contractor	omission variation or payment as abandonment
42(3)	failure by the employer to give possession in accordance with the terms of clause 42	additional cost plus profit on additional work
46(3)	request by the employer or the employer's representative for accelerated completion	terms and conditions as agreed before action is taken
47(5)	repayment of liquidated damages when subsequent extension of time is granted	amount deducted plus interest

Clause	Description	Entitlement
49(3)	work of repair during defects correction period not due to default of the contractor	value of such work as if additional work
50	searches, tests and trials made to determine cause of any defect not found to be contractor's liability	cost of the work carried out
52(1)	valuation of variations agreed prior to the order being given or work commenced	amount agreed
52(2)	valuation of variations not so agreed	quotation or other negotiated amount
52(3)	additional or substituted work ordered in writing by the employer's representative to be on daywork	as daywork provisions in clause 56(3)
58(2)	use of contingencies and prime cost items	valuation in accordance with clause 52 or as the contract otherwise provides
60(7)	interest on overdue payments	compound interest at 2% above base rate
62	urgent repairs ordered in writing by the employer's representative not the contractor's liability	payment implied
63(4)	abandonment of the works due to frustration or war	payment as detailed in clause 63(4)
64(3)	suspension or reduction of work rate arising from failure by the employer's representative to certify or by the employer to pay the amount due	additional cost

Clause	Description	Entitlement
64(6)	termination of the contractor's employment pursuant to the employer's default	payment as detailed in clause 64
65(5)	payment on valuation after termination due to the contractor's default	balance due
69(2)	changes in statutory taxes, levies, contributions etc after the date of tender	net increase or decrease
70(2)	value added tax	tax properly chargeable

17.6 *The amount recoverable under the Design and Construct Conditions*

The D and C Conditions, like other ICE contracts, use a variety of phrases to describe the amount recoverable by the contractor under the various express provisions of the contract. Learned commentators on earlier Conditions have been forceful in their criticism of the multiplicity of phrases so it is disappointing to find that the D and C Conditions have acquired yet another phrase 'any reasonable cost'.

How this phrase fits in with the rules on causation and remoteness of damage discussed above is likely to be a matter of lively debate. There is not the same problem in the D and C Conditions as there is in the ICE Sixth edition of some provisions referring to both recovery of cost and valuation as a variation – methods which do not necessarily result in the same amount. There are still references in some clauses to valuations as a variation in accordance with clause 52 but since clause 52 does not contain any detailed rules for valuations based on bill rates such references would appear to be to quotations or negotiated amounts.

Profit on additional work

The inclusion of profit in certain claims on additional work follows the principle adopted in the ICE Sixth edition.

It is possible that the need for additional work, particularly temporary work, can arise under clauses which refer only to 'cost' – clause 20(3)(c) for example on rectification of loss or damage. The contractor may not be able to recover profit on this work unless he can bring his claim within the scope of some other provision.

Expense of the employer

The phrase 'expense of the employer' used in some clauses has a long history in ICE contracts but its meaning is not defined.

Expense is generally thought to include a reasonable allowance for profit on additional work but there is some doubt on whether profit can also be applied to the delay and disruption costs of the causative event.

17.7 Implied terms for additional payments

The express provisions of the D and C Conditions are not the limit of the contractor's entitlement to additional payments. The right to such payments can also arise out of implied terms founded either on particular provisions of the contract or on common law principles.

Implied from particular provisions

Two obvious examples of the first type are given in the table in section 17.5 above. Thus in clause 26(3)(c) the employer warrants that he has obtained all necessary planning permissions – but the clause says nothing as to payment if he has not. Similarly, clause 62 requires the contractor to carry out urgent repairs as ordered by the employer's representative – but it deals only with the employer's right to recover from the contractor and fails to provide for payment to the contractor in appropriate circumstances.

Less obvious, but potentially just as important, are clauses which state duties or obligations but fail to provide specific remedies for breach. For example, clause 2(1)(a) requires the employer's representative to carry out the duties specified or implied in the contract. Failure to carry out such duties is a breach for which the employer is liable. Another example is clause 11(1) by which the employer is deemed to have made available all information on the site obtained in investigations relevant to the works. Failure to provide such information would leave the employer exposed to claims for breach.

It is not proposed to list all the clauses of the Conditions which could give rise to claims on an implied basis, suffice it to say that where there is a duty there is always the possibility of breach and, from that, the possibility of damage and liability.

Implied from the contract generally

The law allows terms to be implied into a contract to give it business efficacy.

Thus in *Constable Hart & Co. Ltd* v. *Peter Lind & Co. Ltd* (1978) work was carried out on a surfacing contract after the date on which a fixed price quotation expired. It was held that the contract was subject to an implied term that such work would be at a reasonable price.

In *Bacal Construction (Midlands) Ltd* v. *Northampton Development Corporation* (1976), a case often quoted as an object lesson of the dangers of altering a standard form of contract to a design and build contract, ground conditions differed from the information given in borehole details provided by the employer. The foundation designs submitted by the contractor had to be revised with greater cost. It was held that there was an implied term that the ground conditions would accord with the information given and that the contractor could recover his extra costs.

17.8 Meaning of cost

Clause 1(5) of the D and C Conditions gives a comprehensive definition of cost:

> 'The word "cost" when used in the Conditions of Contract means all expenditure including design costs properly incurred or to be incurred whether on or off the Site and overhead finance and other charges (including loss of interest) properly allocatable thereto but does not include any allowance for profit.'

This is much the same definition as used in the ICE Sixth edition. The phrase 'including design costs' is added as also is the bracketed phrase 'including loss of interest'.

Interpretation of the definition is by no means straightforward and there is considerable scope for argument on how it governs remoteness and the important matter of off-site overheads. Much depends on how the phrases 'expenditure – properly incurred' and 'properly allocatable thereto' are taken to apply.

Remoteness of cost

It has been suggested that the words 'all expenditure properly incurred' avoid the rules for remoteness of damage developed from the case of *Hadley* v. *Baxendale* (1854). In other words, if the contractor can show that he has incurred a cost, however remote, that cost can be recovered under the definition in clause 1(5) even though it would not be allowed in a claim for damages. This sounds a doubtful proposition and the answer may lie in the word 'properly' which clearly has some restrictive effect.

Off-site overheads

There are two problems with off-site overheads. Firstly only rarely will the contractor's head office overheads actually increase as a result of a claims situation on a particular contract. If extra cost is the basis for recovery there will rarely, therefore, be anything to recover.

The second problem relates to claims for loss of contribution to overheads – a situation commonly created by delay in completion of the works. Such claims are acceptable at common law as confirmed by His Honour Sir William Stabb QC in *J F Finnegan Ltd* v. *Sheffield City Council* (1988) when saying:

> 'It is generally accepted that, in principle, a contractor who is delayed in completing a contract, due to the default of his employer, may properly have a claim for head office or off-site overheads during the period of delay, on the basis that the workforce, but for the delay, might have had the opportunity of being employed on another contract which would have had the effect of funding overheads during the overrun period.'

However the calculation of the amount due is not only difficult in itself, although various formulae have been devised and accepted by the courts in appropriate circumstances, but the amount is clearly a loss as opposed to a cost.

Other charges properly allocatable

It can be argued that the words 'other charges properly allocatable' excuse the contractor from having to prove expenditure has actually been incurred in respect of overheads. That would permit contractors to continue the long standing practice of adding a percentage to cost based on figures derived from the tender or audited accounts.

Against this it can be said that the words may not be intended to qualify overheads and finance charges but may simply indicate that there may be other classes of expenditure which can be put into cost. This would be more in keeping with the common law position that damages have to be proved.

Expenditure 'to be incurred'

The precise meaning of this phrase is not clear. It can hardly be intended that a contractor should be able to recover cost in advance of its expenditure. Some employers are known to be deleting the phrase from the ICE Sixth edition.

Financing charges

Interest and financing charges seem on the face of it to be the same but in law they are different.

Interest charges relate to late payment of a sum due. Financing charges relate to costs (or loss and expense) which form the subject of claims for the period from which such costs are incurred to the time of application for payment or certification.

The principle concerned in the payment of financing charges is that the contractor has incurred expense in financing the primary cost involved in his claim and this financing charge, therefore, is not interest on a debt but a constituent part of the debt.

The principle was established by the Court of Appeal in the case of *F G Minter Ltd* v. *Welsh Health Technical Services Organisation* (1980).

The decision in the *Minter* case was followed by the Court of Appeal in *Rees and Kirby Ltd* v. *Swansea City Council* (1985) where it was confirmed that calculation of financing charges should be on a compound interest basis.

It is now generally accepted, following the *Minter* and *Rees and*

Kirby cases that, under most standard forms of construction contracts, the contractor is entitled to include in his application for extra cost the financing charges he has incurred up to that time although the Department of the Environment form GC/Works/1 (edition 3) and its Single Stage Design and Build version operate an unusual combined scheme for interest and financing charges.

The D and C Conditions not only accept the principle of financing charges but also clarify a point missing from the ICE Sixth edition by including reference to loss of interest as well as to the cost of borrowed money.

17.9 Procedures and notices

Clause 53 of the D and C Conditions deals specifically with the procedures and notice requirements for claims for additional payment.

The clause is virtually identical to clause 52(4) of the ICE Sixth edition with the employer's representative substituted for the engineer, and with clause 52(4)(a) relating to higher rates and prices omitted.

Requirements of clause 53

The requirements of clause 53 can be briefly summarised as follows:

- The contractor is to give notice of claims as soon as reasonable and in any event within 28 days.
- The contractor is to keep such contemporary records as necessary.
- The employer's representative may instruct the contractor to keep further records.
- The contractor is to permit the employer's representative to inspect his records.
- The contractor is to submit a first interim account.
- The employer's representative may require the contractor to submit up-to-date accounts.
- The contractor is to submit such accounts with accumulated totals.
- The contractor is entitled to payment only to the extent the employer's representative is not prejudiced from investigating the claim.

- The contractor is entitled to interim payment of amount considered due by the employer's representative.
- Where the contractor has not substantiated the whole of the amount claimed he remains entitled to payment of any amount which is substantiated.

Scope of clause 53

Clause 53 covers any claim which is made 'pursuant to any clause of these Conditions'.

Pursuant means in accordance with, or in conformity with, so there is no doubt that all clauses in the Conditions which have a procedure for recovery of extra cost are covered by clause 53. Prominent amongst such clauses are clauses 5, 12, 31 and 40. What is less clear is whether clauses which permit recovery of extra cost or expense but have no procedure or reference to clause 53 are also covered. For example, is the contractor bound by clause 53 if he claims under clause 32 for fossils or under clause 36 for samples? The answer is probably – yes.

However, it is doubtful if the contractor is bound under clause 53 in respect of those clauses which refer to an indemnity.

Extra-contractual claims

The contractor is almost certainly not bound by clause 53 in respect of claims for breach for which no provision is made in the contract – that is, for extra-contractual claims.

Failure to comply with clause 53

Clause 53(4) is a saving provision for the contractor which confirms that if he fails to comply with any of the requirements of clause 53 his entitlement to additional payment is not lost but is only restricted to the extent that the employer's representative has not been prevented from investigating the claim.

Purpose of clause 53

The purpose of clause 53 is not to confer on the employer's

representative any power to finally settle claims for additional payment. The clause relates principally to the contractor's right to interim payment on such claims and to the corresponding safeguards for the employer by way of substantiation of the amounts claimed.

17.10 Differences from the ICE Sixth edition

It may be helpful to readers more familiar with the ICE Sixth edition to have a brief note on the principal differences between the two sets of Conditions in the matter of additional payments.

Express provisions – omissions

Omitted from the D and C Conditions are these provisions in the ICE Sixth edition:

clause 7(4) failure by the engineer to issue additional drawings within a reasonable time

clause 13(3) compliance with engineer's instructions

clause 14(8) delay caused by engineer's consent to contractor's proposed methods or limitations imposed thereon

clause 17(2) cost of rectifying setting-out errors resulting from incorrect data

clause 52(2) refixing of rates rendered unreasonable or inapplicable by any variation

clause 55(2) correction of errors in the bill of quantities

clause 56(2) changes to rates rendered unreasonable or inapplicable by changes in quantities

clause 59(4) recovery of losses following valid termination of a nominated sub-contract

clause 60(8) amounts overpaid in respect of nominated sub-contractors.

Express provisions – additions

Additional to the ICE Sixth edition are these provisions in the D and C Conditions:

clause 6(2) failure by the employer's representative to give consent to designs and drawings within a reasonable time

clause 8(3) failure by the employer's representative to give consent to the quality plan within a reasonable time

clause 64(3) cost of suspending or reducing work rate in consequence of failure in payment

clause 64(6) payment on termination due to the employer's default.

Measurement, variations, instructions

The nature of the D and C Conditions, certainly if priced on a lump sum basis, makes it inevitable that contractors will be deprived of many familiar claims for additional payment.

Measurement, variations and instructions have traditionally been the most fruitful source of claims and these as can be seen from the above and as shown by experience with other design and construct contracts will not have the same impact in the D and C Conditions as in the ICE Sixth edition.

Chapter 18

Measurement, certificates and payments

18.1 Introduction

In this chapter the following clauses of the D and C Conditions are examined:

clause 55 – quantities
clause 56 – measurement and daywork
clause 58 – contingency and prime cost sums
clause 60 – payments and retentions
clause 69 – labour tax fluctuations
clause 70 – value added tax.

Two obvious omissions in the numbering sequence will be noted – clause 57 and clause 59. Neither are used in the D and C Conditions. In the ICE Sixth edition, clause 57 deals with the method of measurement and clause 59 with nominated sub-contractors. Wisely, the D and C Conditions do not attempt to address how the works should be measured; and equally wisely, the Conditions make no provision for nominated sub-contractors.

Quantities

Most contracts under the D and C Conditions will probably be lump sum, so it might be thought that clause 55 on quantities will, in many cases, be superfluous. But that is not quite the case. Clause 55 does not, as it does in the ICE Sixth edition, place any obligations on the employer in respect of inaccuracies or omissions. In the D and C Conditions, clause 55 simply emphasises that quantities, however included in the contract, are estimated only and that errors or omissions do not release the contractor from any of his obligations.

Thus clause 55 in the D and C Conditions is not in any way a claim clause as it is in the ICE Sixth edition. But it is of general application as a statement of the contractor's obligation.

In the event that the D and C Conditions are used with an employer-prepared bill of quantities, interesting contractual questions could arise as to how errors or omissions should be dealt with. Contractors would do well to have the matter resolved before entering into contract.

Measurement

The D and C Conditions strive, quite deliberately, to avoid the label of a remeasurement contract. Consequently there is much in the wording of clause 56(1) on attending for measurement which is of provisional effect. The clause applies only to the extent that it is necessary.

The major change from the ICE Sixth edition is that there is no duty on the employer's representative, as there is on the engineer, to determine by admeasurement the value of the work done.

As with clause 55 this raises the question – what happens if the Conditions are used for a remeasurement contract with a bill of quantities? Clause 56 seems to suggest that either party can take the initiative.

Daywork

The daywork provisions of the D and C Conditions are the same as those in the ICE Sixth edition.

Surprisingly, perhaps, there is a tendency in design and construct contracting to make more use of dayworks than in traditional contracting simply because this is often the least contentious way of valuing variations.

Contingencies

The D and C Conditions make provision for both contingencies and prime cost items to be included in the contract. However, unlike the ICE Sixth edition, which has provisional sums and prime cost items, the Conditions do not attempt to regulate how either should be used.

Thus the D and C Conditions do not link prime cost items with nominated sub-contractors. Nor do the Conditions require orders to be given in respect of the use of contingencies or prime cost items. They simply require consent to commencement of the appropriate work.

Certificates and payment

In traditional contracting, interim payments are usually made on the estimated value of the work executed – with the amount due ascertained by monthly measurement. In design and construct contracting, particularly with lump sum pricing, monthly measurement in the conventional sense is rarely appropriate.

The D and C Conditions, in common with other design and construct forms, describe the contractor's entitlement to interim payments as 'the amounts due under the contract'. It is usually necessary to look outside the conditions of contract, to a payment schedule or similar, to determine how the amount due is calculated. The D and C Conditions follow this approach but also provide a fall-back provision in case there is no payment schedule.

In the matters of final account, retention and entitlement to interest on overdue payments, the D and C Conditions are generally similar to the ICE Sixth edition.

18.2 Clause 55 – quantities

Clause 55(1) – quantities are estimates

Clause 55(1) of the Conditions states that where the contract includes any:

- bill of quantities
- schedule of works, or
- contingencies

the quantities shall be deemed to be the estimated quantities of work and are not to be taken as the estimated or correct quantities of work to be constructed by the contractor in fulfilment of his obligations under the contract.

This reads much the same as clause 55(1) of the ICE Sixth edition but it is of very different effect.

Firstly, in the ICE sixth edition, there will almost certainly be quantities, and those quantities are prepared by or on behalf of, the employer. So clause 55(1) amounts to a statement that the employer does not warrant that his quantities are correct. Under the D and C Conditions, there may not be any quantities, but if there are, they are more likely to have been prepared by the contractor than the employer. So in the D and C Conditions, clause 55(1) is more of a statement that the contractor cannot avoid his obligations by reference to the quantities.

Secondly, in the ICE Sixth edition, clause 55(1) needs to be read in conjunction with clause 55(2) and clause 56(2). Clause 55(2) provides that errors or omissions shall be corrected with corresponding adjustment of the contract price and clause 56(2) provides for increases or decreases in rates should the actual quantities be more or less than those in the bill of quantities. In the D and C Conditions there is nothing on the correction of errors, omissions or adjustment of rates and nothing therefore can be read into clause 55(1) to give the contractor any grounds for claim.

Clause 55(2) – errors and omissions

Clause 55(2) states that no error in description or omission from the bill of quantities, schedule of works or contingency shall vitiate the contract or release the contractor from his obligations under the contract.

This clause is complementary to clause 55(1) in that whilst clause 55(1) deals with the quantities in the numerical sense, clause 55(2) deals with item descriptions for the quantities. But, as with clause 55(1) nothing can be read into clause 55(2) beyond its declaration that the contractor is not excused performance on account of errors or omissions.

Preparation of bills of quantities etc

Clause 55 leaves open the matter as to which party is responsible for the preparation of bills of quantities, schedules of work or contingencies.

Common sense suggests that if the contractor is responsible for preparing the design, he, and he alone, must be responsible for preparing any bill of quantities or schedule of work to be included in the contract. But the D and C Conditions do contemplate that the employer might undertake some of the design himself, before passing on responsibility to the contractor under clause 8(2)(b). So it is not impossible that the employer might produce a bill of quantities.

If such a bill is included in the contract, there is nothing in the Conditions to say that the contract price should be adjusted for errors or omissions. And, since the Conditions do not expressly require remeasurement, there is no certain linkage between clause 55 (quantities) and clause 56 (measurement).

18.3 Clauses 56(1) and 56(2) – measurement

Clauses 56(1) and (2) of the D and C Conditions correspond to clause 56(3) of the ICE Sixth edition. That is, they deal with attending for measurement – but only then to the extent that it is necessary.

The D and C Conditions omit completely:

- any requirement that the value of the works should be determined by admeasurement (as clause 56(1) of the ICE Sixth edition)
- any provision for increases or decreases of rate should the actual quantities be more or less than those stated (as clause 56(2) of the ICE Sixth edition).

Clause 56(1)

Clause 56(1) states that 'If and to the extent that it becomes necessary to measure any part of the Works' then either the contractor or the employer's representative shall give reasonable notice in writing to the other requiring him to attend or assist in making such measurement.

The ICE Sixth edition, which puts a duty of measurement on the engineer confines its wording to the engineer's requirements for the contractor to attend. In itself, the widening of the wording of clause 56(1) in the D and C Conditions to cover the contractor's requirements may seem of little importance but see the comment below on clause 56(2).

Clause 56(2)

Clause 56(2) states that should either the contractor or the employer's representative fail to attend for measurement when so requested, the measurements made by the other shall be taken to be the correct measurement of the works.

Contractors will no doubt welcome this and wish perhaps for a similar provision in the ICE Sixth edition which provides only for measurements made by the engineer to be taken as the correct measurements. And although the change in the D and C Conditions may not be put to use by contractors in every contract, since with lump sum pricing site measurement generally will be of lesser importance, it will still be effective in respect of the measurement of variations, unforeseen conditions and the like.

Accuracy of measurements

The express contractual provision that certain measurements 'shall be taken to be the correct measurement of the Works' suggests that such measurements cannot later be challenged in dispute proceedings. But this may be a simplification since if the measurements are patently wrong they cannot lead to an entitlement to payment for work which has not been undertaken.

The provision is probably only effective in establishing a right to payment, when on the balance of probability, the measurements are of the right order.

18.4 Clause 56(3) – daywork

Employers will need to think carefully about how they wish to regulate daywork payments under the D and C Conditions. If they remain silent on the issue they will find that the contractor is entitled, under clause 56(3), to payment at the rates and prices contained in the Schedules of Dayworks issued by the Federation of Civil Engineering Contractors (FCEC).

The position is the same under the ICE Sixth edition but more often than not contracts are amended to introduce some element of competitive pricing into daywork rates. The results are not always as certain and dispute free as intended but employers should obtain the benefit of lower rates than in the FCEC Schedules if the amendments are competently drafted.

The problem in using the D and C Conditions is that the employer's representative may direct his full attention to technical data in the employer's requirements with reliance on a lump sum price to safeguard the employer's financial interest. Amending the daywork provisions of the Conditions or substituting alternative daywork schedules to the FCEC Schedules can easily be overlooked.

Returns as directed

Clause 56(3) states that daywork returns shall be in the form, and delivered at the times, as the employer's representative 'shall direct'.

Thus a duty is placed on the employer's representative to direct on daywork returns and the traditional requirement for daily records in duplicate does not apply unless it is so directed.

In the event that the employer's representative fails to issue any directions, the contractor's returns whenever, and however, presented would have to be considered.

Returns to be agreed within a reasonable time

Clause 56(3) also states that daywork returns shall be agreed within a reasonable time.

This puts a duty on both parties but the greater burden is perhaps on the employer's representative. The contractor is entitled to include daywork values in his interim applications for payment whether or not they are agreed and failure by the employer's representative to certify leaves the employer liable for interest charges.

Ordering materials

The final provision of clause 56(3) is that before ordering materials the contractor shall 'if so required' submit quotations to the employer's representative for approval.

This is unchanged from the ICE Sixth edition but it may not fully recognise the contractor's design obligations and his right to select the materials he believes to be appropriate.

18.5 Clause 58 – contingency and prime cost items

Definitions

Clause 1(1)(i) defines a 'Prime Cost Item' as an item in the contract which contains a sum referred to as 'Prime Cost' which will be used for the supply of goods, materials or services.

Clause 1(1)(j) defines 'Contingency' as any sum included and so designated in the contract as a specific contingency, which may be used for carrying out work or the supply of goods, materials or services in accordance with the specific requirements.

Clause 58 which deals with the use of contingency and prime cost sums does not in any way expand on these definitions. But note an important distinction in the definitions. Prime cost items 'will' be used; contingencies 'may' be used.

Contingency

The everyday meaning of contingency is an allowance which may or may not be used and the D and C Conditions do not expressly or impliedly suggest any other meaning.

The only restriction applied by the Conditions is that any contingency shall be a 'specific' contingency. This suggests that any 'general' contingency included in the contract is not a 'contingency' within the meaning of the Conditions and is not, therefore, subject to the control provisions of clause 58.

The ICE Sixth edition refers to 'provisional sums' rather than 'contingencies'. There appears to be no significance in this as the definition of a provisional sum is that it is a specific contingency.

Prime cost item

The accepted meaning of prime cost in the construction industry is cost which is reimbursable on the basis of invoices plus a percentage allowance for the contractor's overheads and profit. In many construction contracts, prime cost items are used solely for expenditure on nominated sub-contractors.

In the D and C Conditions it is clearly intended that prime cost items should be valued at invoiced cost plus contractor's percentage since item 4 of part 2 of the appendix to the form of tender reads: 'Percentage adjustment for Prime Cost Items (Clause 58(2))'. However, because the Conditions do not provide for nominated sub-contractors the actual process of determining how the prime cost item is expended is left open. It may be by contractor's choice or it may be that firms or materials are specified in the employer's requirements.

Fixing the amounts

In traditional contracting it is always the responsibility of the engineer/employer to fix the amounts of any prime cost items, contingency sums or provisional sums. It is then the responsibility of the engineer to direct how they should be expended.

In design and construct contracting, the contractor, given the chance, will be only too happy to include prime cost sums, contingency sums or provisional sums in his price. The more he can include, the less risk he carries on price; and the lower the sums included, the more attractive his price seems to be. But, as employers frequently learn to their cost and dismay, such sums are not ceilings on expenditure and the contractor is entitled to recover his actual expenditure, however inadequate his allowances.

With this transfer of risk from the contractor to the employer and the tender advantage gained by inadequate allowances it is advisable that employers take full control of the amounts for all prime cost items, contingency sums or provisional sums and include the same in their employer's requirements.

Clause 58(1) – use of contingency and prime cost items

Clause 58(1) states that the contractor shall not commence work on any contingency or prime cost item without the consent of the employer's representative; which consent shall not be unreasonably withheld.

Unlike the ICE Sixth edition, the clause does not address the detail of how the work should be carried out, nor does it empower the employer's representative to give instructions.

In some cases instructions may be necessary, for example, when the employer has included a prime cost sum or contingency for tree planting. It would be stretching the wording of clause 51 to describe an instruction in such a case on the species of trees required as a variation. It would be no more than clarification of the employer's requirements. Perhaps clause 6 of the Conditions on further information should be seen as applying.

Clause 58(2) – valuation and payment

Clause 58(2) provides that contingencies and prime cost items shall be valued and paid for:

- in accordance with clause 52, or
- as the contract otherwise provides.

It also confirms that the percentage for adjusting prime costs shall be the figure stated in the appendix to the form of tender. This is a figure inserted by the contractor when tendering.

The reference to clause 52 follows that in the ICE Sixth edition but the two situations are not identical. In the ICE Sixth edition most prime cost items are utilised by nominated sub-contracts and the valuation rules of clause 59 apply. And then in the ICE Sixth edition the rules for valuation under clause 52 are very different from those under clause 52 of the D and C Conditions. The latter rules are essentially based on contractor's quotations and there may be some debate on whether they adequately cover valuations which should strictly be on the basis of expenditure incurred.

Programming and delay

There is a further complication with the reference in clause 58(2) to valuation in accordance with clause 52. This is because clause 52 requires quotations to give value and delay consequences.

There have always been arguments in traditional ICE contracts on whether the contractor should include in his programme for prime cost items, contingencies or provisional sums and the answer is probably that he should – at least up to the value of the sums stated. That, of course, is not to deny the contractor the right to an extension of time if he is delayed awaiting instructions or if there is unavoidable delay attached to complying with any instruction.

In the D and C Conditions the same principles should apply but there will certainly be arguments that, because clause 52 expressly refers to delay consequences, any valuation arising from clause 58 should allow for such consequences.

18.6 Clause 60(1) – interim statements

Clause 60(1) of the D and C Conditions provides that the contractor shall submit to the employer's representative 'at such time and in such form' as the contract prescribes statements showing the amounts which, in the opinion of the contractor, are due under the contract.

Compared with the corresponding clause in the ICE Sixth edition which refers to statements at monthly intervals as prescribed in the specification it can be seen that the D and C Conditions are far more flexible. The contract may prescribe periodic payments or stage payments but there is no automatic right from the Conditions alone to monthly payments.

Detail of statements

Clause 60(1) requires the interim statements to show separately where appropriate:

- each amount expended against contingencies and prime cost items
- a list of goods or materials delivered to site but not yet incorporated in the permanent works and their value
- a list of goods or materials not yet delivered to site but of which the property has vested in the employer and their value
- the estimated amounts to which the contractor considers himself entitled in connection with all other matters for which provision is made under the contract.

Again there is a notable difference from the ICE Sixth edition which requires the statement to show principally 'the estimated contract value of the Permanent Works executed up to the end of that month'. Again, the change is made in the interests of flexibility.

This is necessary in the D and C Conditions because the payment schemes for lump sum contracts do not necessarily relate interim payments to the value of work done. And, in any event, in design and construct contracts there is the payment of design fees to consider.

Statement not necessary

As with the ICE Sixth edition a monthly statement need not be submitted if, in the opinion of the contractor, the amount will not justify the issue of an interim certificate.

Goods on-site and off-site

Payment for materials before they are fixed on the employer's land, at which time the employer obtains good title, is always risky. Contracts, by vesting clauses and the like, may purport to give the employer good title but such clauses are only effective as between the employer and the contractor. Receivers, liquidators and suppliers who have not been paid are rarely deterred by such devices.

As to the value of unfixed goods, this should be on the basis of invoiced cost not of the contractor's mark up.

All other matters

Clause 60(1)(e) uses the phrase 'all other matters' to cover the amounts in the contractor's interim statement not relating to contingencies, prime cost items or unfixed goods and materials.

It is not absolutely clear from clause 60(1) whether 'all other matters' includes the contractor's principal amount, however that might be due, or whether it relates only to subsidiary matters such as claims as it does in the ICE Sixth edition. The probability is that the phrase is intended for subsidiary matters.

Temporary works and contractor's equipment

The ICE Sixth edition expressly includes with 'other matters' reference to temporary works and contractor's equipment for which separate amounts are included in the bill of quantities.

The D and C Conditions omit any such reference and it is left to the payment scheme of each contract to indicate how payments for temporary works and contractor's equipment should be made.

Extra-contractual claims

It is doubtful if claims for breach of contract, not expressly covered in the Conditions, should be included as 'other matters' within clause 60.

The wording in clause 60(1)(e) – 'for which provision is made under the Contract' – indicates that only contractual claims are covered by the clause and this suggests that submissions and payments for breach are to be dealt with separately.

18.7 Clause 60(2) – interim payments

The D and C Conditions retain the long standing rule of civil engineering contracts that the contractor should be paid within 28 days of delivery of his statements. But the difference, as noted above, is that the Conditions do not provide expressly for monthly payments.

Clause 60(2) details the actions to be taken by the employer's representative according to whether or not the contract contains a payment schedule.

If there is a payment schedule, clause 60(2)(a) requires the employer's representative to certify:

- when the progress of the works, or any part, has reached the state required for payment of the relevant interim or stage payment
- the amount of such payment less retention.

If there is no payment schedule, clause 60(2)(b) requires the employer's representative to certify:

- the amount which in the opinion of the employer's representative on the basis of the contractor's statement is due to the contractor (less retention)
- such amounts as the employer's representative may consider proper in respect of unfixed goods and materials.

In both cases, the employer's representative is to certify and the employer is to make payment within 28 days of the contractor's application. Late payment entitles the contractor to interest under clause 60(7) and can give the contractor rights to slow down or suspend the work rate under clause 64(1). Prolonged late payment gives the contractor rights of determination under clause 64(4).

Payment schedules

Payment schedules, as discussed in Chapter 5, can be developed for periodic payments, or for stage payments. Periodic payments operate on a pre-determined expenditure plan; stage payments operate on an activity schedule or on other pre-determined events.

The D and C Conditions assume that payment schedules only allow payment when the progress of the works reaches a milestone which permits an interim or stage payment. The wording of clause

60(2)(a) does not cover the situation where there is a payment schedule which links the amount due on a periodic basis to progress against programme. Payments in such a case would seem to fall under clause 60(2)(b) notwithstanding that the clause is stated to apply when there is no schedule.

No payment schedules

The payment scheme in the D and C Conditions when there are no payment schedules is close to that in the ICE Sixth edition. The contractor applies for the amount he considers to be due and the employer's representative certifies the amount which 'in his opinion' is due based on the contractor's statement.

It should be noted that there is no obligation on the employer's representative to certify in the absence of a contractors's statement or to allow in his certificate for matters not included in the statement. The contractor may have his own reasons for undervaluing.

Minimum amount of certificates

Clause 60(3) provides that until substantial completion of the whole of the works the employer's representative is not bound to issue an interim certificate for a sum less than the minimum amount stated in the appendix.

But after the issue of the certificate of substantial completion for the whole of the works, the minimum limits no longer apply and payments are due 'in accordance with the time limits contained in this Clause'.

It is by no means clear what is meant by 'time limits contained in this Clause', unless it means only that the contractor is entitled to be paid within 28 days of any application.

18.8 Clause 60(4) – final account

Clause 60(4) is intended to regulate the timely completion of accounts by requiring the contractor to submit a statement of final account and supporting documentation not later than three months after the date of the defects correction certificate. The statement is to show the value of the works in accordance with the contract together with all further sums which the contractor considers due.

By clause 61(1) there is stated to be only one defects correction certificate and that follows expiration of the last defects correction period.

Final certificate

Within three months after receipt of the final account and verifying information the employer's representative is required to issue a certificate stating the amount which in his opinion is finally due to either the contractor or the employer as the case may be.

Clause 60(4) does not actually use the phrase 'final certificate' and there is no status of finality attached to the certificate as there often is in building contracts where the final certificate is binding unless challenged within a specified time. Moreover since the 'final' certificate covers only sums considered due up to the date of the defects correction certificate, there may well be later certificates as disputes are settled or insurance related matters are resolved.

Credits to the employer

The wording of clause 60(4) leads to some interesting questions on how the employer's representative should handle credits due to the employer.

The clause says that the employer's representative is to certify after giving credit to the employer 'for all sums to which the Employer is entitled under the Contract'. It then says that the amount certified shall 'subject to Clause 47 be paid'. This suggests the amount to be paid is the amount certified after deduction of any liquidated damages for late completion.

This confirms that the employer's representative should not deduct in his final certificate, and certainly not in earlier certificates, for liquidated damages due to the employer for late completion.

18.9 Retention

Clause 60(5)

The wording of clause 60(5) on retention is even more obscure than its counterpart in the ICE Sixth edition. However it simply means that retention is calculated:

- at the rate in the appendix to the form of tender, up to any limit of retention stated therein
- on the amounts due to the contractor
- but it is not applied to any value allowed for unfixed goods or materials
- and it is reduced in respect of any certificates of substantial completion already in issue.

Clause 60(6)

Clause 60(6) covers the release of retention which is due at its simplest in two stages:

- one half upon the issue of any certificate of substantial completion, and
- the balance upon the expiry of the final defects correction period.

Both amounts so due are to be paid within 14 days.

Where there are certificates of completion for sections or parts of the works, the first tranche of retention is released in proportion of the value of work completed to the whole, with the proviso that the amounts so released shall not exceed one half of the limit of retention set out in the appendix. This ensures that the employer has half the retention at substantial completion for the whole of the works.

The second tranche is only released upon the expiry of the last defects correction period and not in instalments where there is more than one such period.

Outstanding claims

Clause 60(6)(c) expressly states that the last tranche of retention money is to be paid upon the expiry of the last defects correction period notwithstanding any outstanding claims by the contractor against the employer. Note this says claims by the contractor; it does not apply to claims by the employer against the contractor.

Outstanding work

However, since it is the expiry of the defects correction period and not the issue of the defects correction certificate which triggers release of the second half of retention money, clause 60(6) sensibly provides that the employer can withhold payment of so much of the retention money as in the opinion of the employer's representative represents the cost of completing any outstanding work.

Retention in trust

There is no express provision in the D and C Conditions for the employer to hold retention monies in a trust fund. Nor, is there such a provision in other standard forms of civil engineering contract.

However in the case of *Wates Construction (London) Ltd* v. *Franthom Property Ltd* (1991) the Court of Appeal decided on a building contract that even in the absence of an express obligation to place retention money into a separate account, the employer was obliged to do so.

It would no doubt be unusual for a contractor in civil engineering to press the employer to establish a separate retention account but it is clear from the case of *MacJordan Construction Ltd* v. *Brookmount Erostin Ltd* (1991) that if the employer's insolvency pre-dates the establishment of such a fund, the contractor is in no better position than any ordinary creditor in respect of any monies held as retention.

18.10 Interest on overdue payments

The D and C Conditions have adopted the same rules for the payment of interest on overdue payments as the ICE Sixth edition. However in the light of the decision in the case of *The Secretary of State for Transport* v. *Birse-Farr Joint Venture* (1993) the provisions of the ICE Fifth edition would be more appropriate. This point is discussed below.

Interest calculations

The scheme in clause 60(7) for the calculation of interest can be summarised as follows:

- interest is to be compounded monthly for each day on which payment is overdue

- payment is overdue from the date it should have been paid (if certified) or the date it should have been certified and paid (if not certified)
- the interest rate is 2% above the base lending rate of the bank specified in the appendix
- interest will be added each month on interest already due to be paid.

Payment of interest

Clause 60(7) makes no provision for the employer's representative to certify interest on overdue payments. It is the responsibility of the employer to pay interest directly it becomes due.

However, the contractor would be wise to include any interest due in his monthly application under clause 60(1)(e) as being an amount for which provision is made under the contract. The employer's representative would then be obliged to certify such interest under clause 60(2).

Findings of an arbitrator

Clause 60(7) provides that if an arbitrator finds that a sum should have been certified by a particular date but was not so certified, it shall be regarded as overdue for payment from either:

- 28 days after the date the arbitrator holds the employer's representative should have certified, or
- from the date of the certificate of substantial completion of the whole of the works where no certifying date is identified by the arbitrator.

This provision is consistent with, and follows, the judgment in the case of *Morgan Grenfell (Local Authority Finance) Ltd* v. *Seven Seas Dredging Ltd* (1990).

In that case under the ICE Fifth edition the arbitrator found in favour of the contractor for unforeseen work in dredging for the Port of Sunderland. The contractor was awarded £1 954 811 plus interest. The arbitrator awarded interest of £967 604 calculated at 2% above bank rate under clause 60(6). If the arbitrator had calculated interest under section 19(A) of the Arbitration Act 1950 the award of interest would have been £187 449 less.

The employer appealed and the court had to decide whether the contractor was entitled to interest under clause 60(6) in respect of amounts included in a statement under clause 60(1) but not certified by the engineer, when an arbitrator later decides they should have been certified. The appeal was dismissed with the court finding that interest under clause 60(6) was allowable.

Meaning of failure to certify

At the heart of the *Morgan Grenfell* case was the long argued point of whether an engineer, in rejecting all or part of a contractor's statement, leaves the employer open to liability for interest if he (the engineer) or an arbitrator subsequently revises upwards the evaluation of the contractor's claim. A common view was that providing the engineer acted in good faith in forming 'his opinion' under clause 60(2), there was no 'failure to certify'. Judge Newey QC in the *Morgan Grenfell* case seemed to go against this in finding in favour of the contractor when saying:

> 'I have reached the same conclusions as the learned arbitrator. In my judgment clause 60(2) requires the engineer, who has been provided with a statement in accordance with clause 60(1), to "certify" the amount which the contractor should be paid. Obviously, that amount should be the amount, which in his "opinion" the contractor should be paid. If the word "opinion" had not been used in the sub-clause some similar word or words would have had to have been used. For the sub-clause to work the engineer has to decide how much the contractor should receive and that is a matter of "opinion", "judgment", "conclusion" or the like. If the engineer certifies an amount which is less than it should have been, the contractor is deprived of money on which he could have earned money.'

It had been widely expected that the *Morgan Grenfell* case would go on to the Court of Appeal but this never happened and it was left to the case of *The Secretary of State for Transport* v. *Birse-Farr Joint Venture* (1993) to re-examine how the provisions for interest in the ICE Fifth edition should apply.

The court in that case held that an engineer will only have failed to certify if:

- he fails to certify altogether
- he under-certifies on an erroneous view of the contract.

The judge said:

> 'The Contractor must demonstrate not merely that a sum which has later been allowed by the Engineer or the Arbitrator was not included in an earlier certificate but must show that there was a failure by the Engineer to perform his duties under the Contract in deciding what sum to certify.'

The court also held that:

- compound interest is not due under the ICE Fifth edition; but only interest on arrears of interest
- until completion the contractor can claim interest on arrears of interest; but thereafter only simple interest accrues.

The *Birse-Farr* decision will have some impact on the D and C Conditions in regard to the meaning of failure to certify but it will not affect the express entitlement in the Conditions to compound interest.

Extra-contractual claims

It is worth noting that the interest provisions of clause 60(7) apply only to amounts due under the contract. The contractor therefore has no contractual entitlement to interest for late certification or payment of an extra-contractual claim.

Interest not financing charges

The interest in clause 60(7) should not be confused with the finance charge in 'cost' as defined in clause 1(5). Interest applies to late payment on a contractor's application; finance charges are part of the cost included in such an application.

18.11 Correction and withholding of certificates

Clause 60(8) empowers the employer's representative to omit from any certificate the value of work done or goods supplied with which he is dissatisfied. For that purpose, or for any other reason which 'may seem proper' the employer's representative may, in any certificate, delete, correct or modify any previous sum certified.

Dissatisfaction

The provision on dissatisfaction recognises that as the works proceed matters may come to light which diminish or possibly eliminate the value of work already paid for in that reconstruction or demolition may be necessary at the contractor's own cost.

Other reasons

There may be some uncertainty on what is meant by 'any other reason which to him may seem proper'.

It is not thought to be a general power for the employer's representative to coerce or punish the contractor for contractual failings and it is most unlikely that it could apply to matters outside the contract – such as knowledge of the contractor's financial difficulties.

Possibly it relates to the withholding from interim certificates, of amounts due under the contract to the employer from the contractor.

Perhaps the provision means not much more than that the employer's representative can correct mistakes or change his mind.

18.12 *Copy certificates and payment advice*

Copy certificates

Clause 60(9) provides that every certificate issued by the employer's representative pursuant to clause 60 shall be sent to the employer and at the same time copied to the contractor 'with such detailed explanation as may be necessary'.

In practice it is usual for the employer's representative to return to the contractor a copy of the application for payment showing any reductions or amendments.

Payment advice

Clause 60(10) applies when the employer pays an amount different from that certified by the employer's representative.

The employer is to notify the contractor 'forthwith' with full details showing how the amount paid has been calculated.

This is a useful addition and a necessary one in the light of clause 47 on liquidated damages which has no conditions precedent to the deduction of damages.

18.13 Clause 69 – labour tax fluctuations

Clause 69 on labour tax matters allocates the risk for any changes post tender to the employer.

Rates in the tender

Clause 69(1) states that the rates and prices in the contract are deemed to take account of the levels and incidence, at the date of return of tenders, of:

- taxes
- levies
- contributions
- premiums or refunds

payable by law by the contractor and his sub-contractors in respect of their employees engaged on the contract.

Moreover, the rates and prices do not take account of matters known at the date of tender to have later effect.

Adjustment to the contract price

Clause 69(2) provides that changes in labour tax matters after the date of tender resulting in a net increase or decrease shall be taken into account in the contract price.

Application of clause 69

Clause 69 applies unless it is deleted from the contract. In the ICE Sixth edition which has an identical clause it frequently is deleted and it can be expected that under the D and C Conditions deletion will be commonplace.

Supporting information

One practical reason for the unpopularity of the clause is that production of the supporting information is a difficult task and the task of checking the information is even worse.

18.14 Clause 70 – value added tax

Clause 70 states that the contractor is deemed not to have allowed for value added tax in his tender.

That does not mean that the contractor is relieved of his responsibility to the Customs and Excise for any value added tax which may be due – he most certainly is not. What the clause does is establish, as between the contractor and the employer:

- that the contract price is exclusive of VAT
- that the employer will refund to the contractor any VAT payable in respect of the contract.

Certificates net of VAT

For the avoidance of doubt clause 70(2) expressly states that all payment certificates under clause 60 shall be net of VAT.

Disputes

Clause 70(3) requires the parties to assist each in the event of dispute with the Customs and Excise – a not uncommon occurrence.

Clause 70(4) states that clause 66 does not apply to disputes on VAT.

Chapter 19

Defaults, determination and frustration

19.1 Introduction

The D and C Conditions cover the subjects of defaults, determination, frustration and war in more logical order and with better balance between the parties than the ICE Sixth edition or other ICE contracts.

To achieve this there has been a considerable amount of rearrangement, and redrafting of the text of the ICE Sixth edition and the insertion of new provisions not previously seen in ICE contracts.

From the perspective of the ICE Sixth edition the position is as follows:

ICE Sixth edition	**D and C Conditions**
• clause 63 – determination for contractor's default	• clause 65
• clause 64 – frustration	• clause 63(1) and clauses 63(3) to 63(6)
• clause 65 – war clause	• clauses 63(2) to 63(6)

The order as the various provisions appear in the D and C Conditions is now:

D and C Conditions	**ICE Sixth edition**
• clause 63(1) – frustration	• clause 64
• clauses 63(2) to 63(6) – war clause and abandonment	• clause 65
• clauses 64(1) to 64(3) – failure to pay the contractor	• no equivalent

• clauses 64(4) to 64(6) – default of the employer	• no equivalent
• clause 65 – default of the contractor	• clause 63

Changes of note

The most important changes in the D and C Conditions are:

- the introduction of reciprocal rights of determination entitling the contractor to determine his own employment in the event of employer's default
- the introduction of provisions entitling the contractor to slow down or suspend work in the event of non-payment
- redrafting of the frustration clause to include the circumstances of its operation.

None of these changes could be said to be particularly related to, or in consequence of, the transfer of design responsibility between the ICE Sixth edition and the D and C Conditions; and the ICE Sixth edition would certainly be improved itself by the incorporation of the changes.

Characteristics of design and construct contracting

In so far that design and construct contracting may have special characteristics applying to such matters as frustration, defaults and determination these are likely to be no more than an increased sensitivity to practical matters such as impossibility and failing to proceed with due diligence. That is, the transfer of design responsibility to the contractor effectively increases the scope of his defaults and potentially opens up new arguments on impossibility.

19.2 Frustration

Clause 63(1) of the D and C Conditions has the marginal note 'Frustration'. The clause reads as follows:

> 'If any circumstance outside the control of both parties arises during the currency of the Contract which renders it impossible

> or illegal for either party to fulfil his contractual obligations the Works shall be deemed to be abandoned upon the service by one party upon the other of written notice to that effect.'

In the ICE Sixth edition, clause 64 is headed 'Frustration' but it carries the marginal note 'Payment in the event of frustration'. That clause reads:

> 'In the event of the Contract being frustrated whether by war or by any other supervening event which may occur independently of the will of the parties the sum payable by the Employer to the Contractor in respect of the work executed shall be the same as that which would have been payable under Clause 65(5) if the Contract had been determined by the Employer under Clause 65.'

It can be seen that there is a significant difference between these two clauses. The clause in the D and C Conditions is, in effect, defining frustration, whereas the clause in the ICE Sixth edition is merely stating what payment should be made in the event of frustration.

Frustration generally

At common law a contract is discharged and further performance excused if supervening events make the contract illegal or impossible or render its performance commercially sterile. Such discharge is known as frustration. A plea of frustration acts as a defence to a charge of breach of contract.

In order to be relied on, the events said to have caused frustration must be:

- unforeseen
- unprovided for in the contract
- outside the control of the parties
- beyond the fault of the party claiming frustration as a defence.

In the case of *Davis Contractors* v. *Fareham UDC* (1956) Lord Radcliffe said:

> 'Frustration occurs whenever the law recognises that without default of either party a contractual obligation has become incapable of being performed because the circumstances in

which performance is called for would render it a thing radically different from that which was undertaken by the contract. *Non haec in foedera veni.* It was not this that I promised to do.'

In that case a contract to build 78 houses in eight months took 22 months to complete due to labour shortages. The contractor claimed the contract had been frustrated and he was entitled to be reimbursed on a quantum meruit basis for the cost incurred. The House of Lords held the contract had not been frustrated but was merely more onerous than had been expected.

Frustration in construction contracts

Frustration, in the true legal sense of a radical change of obligation, is uncommon in construction contracts.

One of the few recorded cases in the UK is *Metropolitan Water Board* v. *Dick Kerr & Co. Ltd* (1918) where the onset of the First World War led to a two year interruption of progress. It was held that the event was beyond the contemplation of the parties at the time they made the contract and the contractor was entitled to treat the contract as at an end.

More recently in a Hong Kong case, *Wong Lai Ying* v. *Chinachem Investment Co. Ltd* (1979), a landslip which obliterated the site of building works was held by the Privy Council to be a frustrating event.

A brief flurry of excitement was caused in the minds of contractors and claims consultants in 1990 when the judge at first instance in the case of *McAlpine Humberoak Ltd* v. *McDermott International Inc* held that a contract had been frustrated by the increased number of drawings issued. But the Court of Appeal in 1992 firmly rejected this. Lord Justice Lloyd said:

> 'If we were to uphold the judge's finding of frustration, this would be the first contract to have been frustrated by reason of matters which had not only occurred before the contract was signed, and were not only well known to the parties, but had also been provided for in the contract itself.'

Payment in the event of frustration

The rules governing the transfer of monies from one party to the other in the event of frustration are exceedingly complex unless the contract itself provides how losses are to fall.

Hence the purpose of clause 64 in the ICE Sixth edition is not to regulate a code of conduct or define frustrating circumstances. It is simply to ensure that the contractor is paid for the work he has done and is not forced by the Law Reform (Frustrated Contracts) Act 1943 into proving that the employer has received valuable benefit.

In the D and C Conditions the scheme for payment on abandonment is set out in clause 63(4).

Frustration in the Design and Construct Conditions

In the D and C Conditions the purpose of clause 63(1) is to define those circumstances which, should they arise, would lead to the contract being deemed to be abandoned.

Under clause 63(1) any such circumstance must:

- be outside the control of both parties
- arise during the currency of the contract (that is – be a supervening event)
- render impossible or illegal the fulfilment of contractual obligations by either party.

Clause 63(1) further requires written notice to be served by one party on the other that abandonment is deemed to have occurred.

The question of whether the scope of clause 63(1) is wider or narrower than common law frustration is best left to the lawyers but two points may be worth brief comment:

- The phrase 'fulfil his contractual obligations' in clause 63(1) looks potentially wider than the common law test of performance. Perhaps it could be argued that if any, rather than the major, contractual obligations are impeded the contract is deemed to be abandoned.
- The contractor's general obligation under clause 8(1) is to design, construct and complete the works 'save in so far as it is legally or physically impossible'. This would appear to apply to pre-existing as well as to supervening events. The Conditions may, therefore, have made physical or legal impossibility a defence to non-performance on a much broader scale than would be implied by the doctrine of frustration.

19.3 Termination in the event of war

The provisions in the D and C Conditions for termination in the event of war are thankfully shortened from those in traditional ICE contracts.

Clause 63(2)(a) provides that if there is an outbreak of war the contractor is to continue to execute the works as far as possible for a period of 28 days.

Clause 63(2)(b) provides that if substantial completion is not achieved within the period of 28 days the works are deemed to have been abandoned unless the parties otherwise agree.

19.4 Effects of abandonment

Clauses 63(3) to 63(6) of the D and C Conditions deal with the effects of abandonment whether arising from frustration or war.

Removal of contractor's equipment

Clause 63(3) requires the contractor to remove all contractor's equipment from the site with all reasonable dispatch.

In the event of the contractor failing to do so the employer is given powers to dispose of the contractor's equipment and to retain his costs before accounting for the proceeds.

Payment on abandonment

Clause 63(4) provides that the contractor is to be paid on abandonment:

- the value of all work carried out
- proportionate amounts in respect of preliminaries
- costs of goods and materials for which the contractor is legally obliged to accept delivery
- any expenditure reasonably incurred in expectation of completing the whole of the works
- the reasonable cost of removing contractor's equipment.

Clause 63(4) further provides that the timescale for finalisation of accounts in clause 60(4) applies as if the date of abandonment was the date of 'issue' of the defects correction certificate.

The word 'issue' may have been inserted in the clause unintentionally because clause 60(4) itself uses the phrase 'the date of the Defects Correction Certificate'.

Works substantially completed

Clause 63(5) deals with allowances for remedying defects and also the release of retention money.

The contractor may at his discretion allow a sum to be set off against the amount due to him on abandonment in lieu of his obligations in respect of making good defects.

The employer is not entitled to withhold payment of retention money except as part of any sum agreed as above.

Contract to continue in force

Clause 63(6) states that, save as the earlier provisions of clause 63, the contract shall continue to have full force and effect.

This would seem to apply principally to the disputes resolution procedures of clause 66. It is difficult to see how much else of the contract can contrive to have full force and effect after abandonment.

19.5 Failure to pay the contractor

It is a fault of most ICE contracts that the contractor is given no contractual remedy for failure by the engineer to certify or the employer to make payment other than an entitlement to interest on sums paid late. But if the contractor's cash flow is disrupted the expectation of interest at some date in the future may be far from adequate in allowing him to continue to fulfil his obligations under the contract.

The D and C Conditions have rectified this fault in clauses 64(1) to 64(3).

Suspension or reduction of work rate

Clause 64(1) provides that if the employer's representative fails to certify or the employer fails to pay within 21 days of the due date then the contractor may, after giving seven days notice in writing to the employer, either:

- suspend work, or
- reduce the rate of work.

Without a contractual right to suspend or slow down the contractor's position is legally uncertain. Thus, in the case of *Canterbury Pipe Lines Ltd* v. *Christchurch Drainage Board* (1979), in a dispute over amounts certified by the engineer on a drainage contract, the contractor suspended work which led to the employer taking the work out of the contractor's hands. It was held that the contractor had no right to suspend. However, in *Hill (J M) & Sons Ltd* v. *London Borough of Camden* (1980), where a contractor, unable to obtain prompt payment, cut his staff and labour, it was held that his action did not entitle the employer to determine the contract.

But what these cases, and others, illustrate is that disputes over alleged late payment often go deeper than the timing of payments – there is more than likely to be dispute over the amount of the payments. So even under the D and C Conditions the contractor's position is not entirely certain.

The contractor may believe that the employer's representative has under-certified (and the meaning of this has now been clarified in the case of *Secretary of State for Transport* v. *Birse-Farr* discussed in Chapter 18) and suspend or slow down under clause 64(1). The employer may dispute that there has been any under-certification. And whilst all will be resolved eventually in conciliation or mediation – what is the position in the meantime? Does the contractor risk determination under clause 65 if he has misjudged his entitlement to suspend or slow down under clause 64(1)? The answer is probably – yes.

Resumption of work

Clause 64(2) states that upon payment by the employer of the amount due, including interest, the contractor shall resume normal working as soon as is reasonably possible.

Delay and extra cost

Clause 64(3) requires the employer's representative to determine:

- any extension of time to which the contractor is entitled
- the amount of additional cost arising as a result of the contractor having suspended or slowed down the work rate under clause 64(1).

Clearly, if there is a dispute on whether the contractor's action is in order clause 64(3) will only be applied with some difficulty. But even if there is no dispute on the contractor's action there may be some disagreement on extension of time.

The contractor may seek to argue, relying perhaps on the decision in the case of *Fernbrook Trading Company Ltd* v. *Taggart* (1979), that he does not want an extension of time and that time is at large. To which the employer will reply that clause 44(1)(b) allows for an extension of time for 'any cause of delay referred to in these Conditions'. On this, the employer is probably right.

19.6 Defaults and determination

The D and C Conditions, unlike the ICE Sixth edition, do not specifically label any clause 'determination'. But clauses 64(4) to 64(6) which deal with employer's defaults and clause 65 which deals with contractor's defaults are provisions for termination of the contractor's employment which amount to determination.

Determination generally

The ordinary remedy for breach of contract is damages but there are circumstances in which the breach not only gives a right to damages but also entitles the innocent party to consider himself discharged from further performance.

Repudiation

Repudiation is an act or omission by one party which indicates that he does not intend to fulfil his obligations under the contract. In civil engineering works, a contractor who abandons the site or an employer who refuses to give possession of the site are examples of a party in repudiation.

Determination at common law

When there has been repudiation or a serious breach which goes to the heart of the contract so that it is sometimes called 'fundamental' breach, common law allows the innocent party to accept the repudiation or the fundamental breach as grounds for determination of the contract. The innocent party would then normally sue for damages on the contract which had been determined.

The problems with common law determination are that it is valid only in extreme circumstances and it can readily be challenged.

Determination under contractual provisions

To extend and clarify the circumstances under which determination can validly be made and to regulate the procedures to be adopted, most standard forms of construction contract include provisions for determination. Building forms have traditionally given express rights of determination to both employer and contractor but civil engineering forms, until the D and C Conditions, usually have given express rights only to the employer.

Many of the grounds for determination in standard forms are not effective for determination at common law. Thus failure by the contractor to proceed with due diligence and failure to remove defective work are often to be found in contracts as grounds for determination by the employer. But at common law neither of these will ordinarily be a breach of contract at all since the contractor's obligation is only to finish on time and to have the finished work in satisfactory condition by that time.

The commonest and the most widely used express provisions for determination relate to financial failures. Again at common law many of these are ineffective and even as express provisions they are often challenged as ineffective by legal successors of failed companies.

The very fact that grounds for determination under contractual provisions are wider than at common law leads to its own difficulties. A party is more likely to embark on a course of action when he sees his rights expressly stated than when he has to rely on common law rights. This itself can be an encouragement to error. Some of the best known legal cases on determination concern determinations made under express provisions but found on the facts to be lacking in validity

In *Lubenham Fidelities* v. *South Pembrokeshire District Council* (1986)

the contractor determined for alleged non-payment whilst the employer concurrently determined for failure to proceed regularly and diligently. On the facts, the contractor's determination was held to be invalid. But in *Hill & Sons Ltd* v. *London Borough of Camden* (1980), with a similar scenario, it was held on the facts that the contractor had validly determined.

Determination of the contractor's employment

Provisions for determination in construction contracts are usually drafted as provisions for determination of the contractor's employment.

The purpose of this is to emphasise that the contract itself remains in existence and that its various secondary provisions remain operative after determination – not least those in the determination clause itself.

Parallel rights of determination

Some construction contracts expressly state that their provisions, including those of determination, are without prejudice to any other rights the parties may possess. That is, the parties have parallel rights – those under the contract and those at common law – and they may elect to use either.

ICE Conditions do not have such a stated alternative but the omission is not significant. The general rule is that common law rights can only be excluded by express terms. Contractual provisions, even though comprehensively drafted, do not imply exclusion of common law rights.

The point came up in the case of *Architectural Installation Services Ltd* v. *James Gibbons Windows Ltd* (1989) where it was held that while a notice of determination did not validly meet the timing requirements of the contractual provisions, nevertheless there had been a valid determination at common law.

Legal alternatives

To take advantage of the parallel rights of determination and to avoid as far as possible the danger that one course of action might be found to be defective, determination notices are sometimes served under both the contract and at common law as legal alternatives.

But on that is must be said that contractual rights which are not also common law rights attract only contractual remedies. So a defective notice under a contractual provision which has no common law equivalent cannot be salvaged by a common law action.

19.7 *Default of the employer under the Design and Construct Conditions*

Clause 64(4) specifies the defaults of the employer which entitle the contractor to terminate his own employment under the contract.

In so far that it gives the contractor reciprocal rights to those long enjoyed by employers under ICE contracts, clause 64(4) must be welcomed. But since its use will, in many cases, mean that the employer has run out of funds, or is otherwise insolvent, clause 64(4) is never likely to be applied with enthusiasm.

Specified defaults

The defaults specified in clause 64(4) are:

- failure by the employer's representative to certify or the employer to pay the amount due to the contractor within 56 days after expiry of the time allowed in clause 60 (28 days)
- assignment or attempted assignment by the employer of the contract, or any part thereof, or any interest or benefit thereunder without the prior written approval of the contractor
- financial failures of the kind:
  - becoming bankrupt
  - being subject to a receiving order or administration order
  - making an arrangement or assignment in favour of creditors
  - agreeing to a committee of inspection
  - going into liquidation, receivership or administration
  - having an execution levied on goods which is not stayed or discharged within 28 days.

Action by the contractor

In the event of a specified default, the contractor may, under clause 64(4):

- give notice in writing to the employer specifying the default
- after seven days notice terminate his own employment under the contract
- extend the period of notice to give the employer an opportunity to remedy his default.

Purpose of the notice period

It is a matter of some legal importance as to whether the notice given by the contractor is simply a notice given of his intention to vacate the site or whether it is a warning period for the employer to remedy his default. Similar questions arise under traditional ICE contracts in respect of notices given to contractors by employers. And, of course, the same applies to notices of contractor's default under the D and C Conditions.

The wording of the D and C Conditions, like that of the ICE Sixth edition, leans to the probability that the notice period should be seen as a warning period. That however leads to difficult questions as to who is to judge whether the default has been remedied and what procedure is to be followed if there is a dispute on whether it has or it has not been remedied.

To say that these are difficult issues is an understatement. In the case of *Attorney General of Hong Kong* v. *Ko Hon Mau* (1988), on conditions similar to the ICE Fifth edition, both parties served notice of determination. The court held that the notices were provisional in the sense that their final validity would not be determined until arbitration, but pending the arbitration, providing they were given in good faith, they were both effective.

Removal of contractor's equipment

Clause 64(5) states that upon the expiry of the seven days notice period, the contractor shall, with all reasonable despatch, remove from the site all contractor's equipment.

This applies notwithstanding the provisions of clause 54 which debars the removal of any of the contractor's equipment without the consent of the employer's representative.

The wording of clause 64(5) is slightly odd in that it says – 'shall ... remove'. If taken in their usual mandatory sense these words would put an obligation on the contractor to remove his equipment after seven days whether or not he intended to extend the period of notice to give the employer more time to remedy his defect.

Payment upon termination

Clause 64(6) provides that upon termination of the contractor's employment pursuant to the clause 64(4) (employer's default) the employer shall be under the same obligations as to payment as if the works had been abandoned under clause 63 (frustration). And, in addition, the contractor is entitled to:

- any amount due to the contractor under clause 63(3) in respect of delay or extra cost arising from late payment
- the amount of any loss or damage arising from the termination.

In total, therefore, the contractor is entitled to:

- the value of all work carried out
- proportionate amounts in respect of preliminaries
- costs of goods and materials the contractor is legally obliged to accept
- any expenditure reasonably incurred in expectation of completing the works
- the reasonable cost of removing contractor's equipment
- amounts due in respect of late payment
- loss or damage arising from the termination.

The final heading 'loss or damage' would include loss of overheads and profit on the uncompleted work in accordance with the usual principles. So in theory, on termination due to the employer's default, the contractor should receive not only the value of work done plus costs incurred but also lost profit.

In practice, of course, the contractor may receive little or nothing if the default is financial failure of the employer and the contractor is only an unsecured creditor.

19.8 Default of the contractor under the Design and Construct Conditions

The provisions in clause 65 of the D and C Conditions for termination by the employer in the event of the contractor's default are virtually identical to those in clause 63 of the ICE Sixth edition.

The only change of any significance is that the D and C Conditions contain an additional default – breach of clause 4(1). That is, sub-contracting the whole of the works without the prior written consent of the employer. The omission of this from the ICE Sixth edition was probably a drafting slip.

Specified defaults

The specified defaults of the contractor under clause 65 fall into four categories:

- assignment without consent; clause 65(1)(a)
- sub-contracting the whole of the works without consent; clause 65(1)(b)
- financial failures; clauses 65(1)(c) and 65(1)(d)
- performance failures; clauses 65(1)(e) to 65(1)(j).

Financial failures

The financial failures are the same as those specified for the employer under clause 64(4)(c):

- becoming bankrupt
- being subject to a receiving order or an administration order
- making an arrangement or assignment in favour of creditors
- agreeing to a committee of inspection
- having an execution levied which is not stayed or discharged within 28 days.

Failures of performance

Under clause 65(1) the specified defaults are:

- abandoning the contract without due cause
- failing to commence without reasonable excuse
- suspending progress without due cause for 14 days after notice to proceed
- failing to remove or replace defective work, goods or materials after notice to do so
- failing to proceed with due diligence despite previous warnings

- being persistently or fundamentally in breach of obligations under the contract.

'Abandoned the Contract'

Abandonment may be patently obvious where the contractor has left the site or has otherwise given notice of his intention not to fulfil his obligations. Such abandonment is repudiation and grounds for common law determination.

However, there can also be situations where the contractor's conduct in ceasing work is neither abandonment nor repudiation. In *Hill* v. *Camden* (1980), the contractor, unable to obtain prompt payment cut his staff and labour. The employer took this action as repudiatory and served notice of determination. Lord Justice Lawton had this to say of the contractor's action:

> 'The plaintiffs did not abandon the site at all; they maintained on it their supervisory staff and they did nothing to encourage the nominated sub-contractors to leave. They also maintained the arrangements which they had previously made for the provision of canteen facilities and proper insurance cover for those working on the site.'

A point worth noting is that the D and C Conditions add the phrase 'without due cause' to clause 65(1)(e). This is not in the ICE Sixth edition. The explanation may lie in clause 63 of the D and C Conditions which uses the word 'abandoned' in connection with frustration.

'Failed to commence'

By clause 41(2) the contractor shall start the works on or as soon as is reasonably practicable after the works commencement date. The default under clause 65 is failing to commence in accordance with clause 41 without reasonable excuse.

Failure to commence within a reasonable time is not of itself a repudiatory breach. The contractor might say that he can complete in a fraction of the time allowed and has every intention of finishing on time. And in design and construct contracting there is plenty of scope for argument on how much time the contractor needs off-site to prepare himself for starting the construction work.

In the D and C Conditions this default needs to be handled with extreme caution.

Suspension of progress

In clause 65(1)(g) the D and C Conditions again include the phrase 'without due cause'. This time because clause 64(1) expressly permits to the contractor to suspend for late payment.

Failure to remove defective work, goods or materials

At common law, failure by the contractor to remove or replace defective work, goods or materials is not, of itself, a breach of contract. It is only a breach of contract if the contractor has failed to do so by the completion date. In *Kaye Ltd* v. *Hosier & Dickinson Ltd* (1972) Lord Diplock said:

> 'During the construction period it may, and generally will, occur that from time to time some part of the works done by the contractor does not initially conform with the terms of the contract either because it is not in accordance with the contract drawings or the contract bills or because the quality of the workmanship or materials is below the standard required by condition 6(1)... Upon a legalistic analysis it might be argued that the temporary disconformity of any part of the works with the requirements of the contract, even though remedied before the end of the agreed construction period, constituted a breach of contract for which nominal damages would be recoverable. I do not think that makes business sense. Provided that the contractor puts it right timeously I do not think that the parties intended that any temporary disconformity should of itself amount to a breach of contract by the contractor.'

The D and C Conditions do, however, in clause 39, empower the employer's representative to order the removal of unsatisfactory work and materials and failure to comply with such an instruction is a breach of contract. The remedy of determination in clause 65 is extreme and perhaps the mere serving of notice under clause 65 will usually be effective in getting action from the contractor.

Prior to the case of *Tara Civil Engineering* v. *Moorfield Developments Ltd* (1989) it was generally thought that under the ICE Fifth edition the engineer could not serve notice under clause 63 until he had acted under clause 39. However, the judge in the case did not accept that proposition. He said:

> 'I am in no doubt at all that clause 63 can and should be construed without any suggestion that it is limited by clause 39 or that it should be preceded by a notice which is in some way identifiably referable to clause 39. The engineer and the employers have various options open to them under the contract and those options should not be restricted by the sort of argument that has been put in this case.
>
> I therefore find that the engineer has issued documents which on their face appear to put in motion the machinery of clause 63.'

The same principle will apply to the relationship between clause 65 and clause 39 in the D and C Conditions.

Failing to proceed with due diligence

This is a matter which should be approached with the greatest of caution. The courts have been most reluctant to impose on contractors any greater obligation than to finish on time.

In *Greater London Council* v. *Cleveland Bridge* (1986) the Court of Appeal refused to imply a term into a building contract that the contractor should proceed with due diligence notwithstanding the inclusion of that phrase in the determination clause of that contract. The point was repeatedly made in that case that the contractor should be free to programme his work as he thought fit. For other cases showing the difficulty of defining due diligence and failure to proceed with it, see *Hill* v. *Camden* (1980) and *Hounslow* v. *Twickenham Garden Developments* (1971)

Failure by the contractor to proceed in accordance with his approved clause 14 programme might provide some evidence to support a charge of failing to proceed with due diligence although failure to proceed to the programme is not itself a breach of contract.

The reference to 'previous warnings' in clause 65(1)(j) should, perhaps, be read in conjunction with clause 46. That clause places a duty on the employer's representative to notify the contractor if he considers progress too slow to achieve completion by the due date.

Persistently or fundamentally in breach

The defaults of 'persistently or fundamentally in breach of his obligations under the Contract' may be mutually exclusive; that is, they may be persistent breaches of minor obligations or a single breach which goes to the heart of the contract. It is not clear whether the 'previous warnings' apply to these defaults as well as to failing to proceed with due diligence.

Action by the employer

Clause 65(1) entitles the employer to expel the contractor from the site after giving seven days notice in writing specifying the contractor's default.

As in clause 64, the period of notice may be extended. But this time the extension is allowed by the employer to give the contractor the opportunity to remedy his default.

Clause 65(1) further provides that where a notice of termination is given pursuant to a certificate issued by the employer's representative, such notice shall be given 'as soon as is reasonably possible after receipt of the certificate'.

This should avoid the situation in the case of *Mvita Construction* v. *Tanzania Harbours Authority* (1988), under FIDIC Conditions which matched those of the ICE Fifth edition. There, the employer took three months before acting on an engineer's certificate. It was held that the employer was bound to give his notice within a reasonable time after the engineer's certificate to avoid a change in circumstances.

Employer's representative to certify in writing

For the performance defaults in clause 65(1)(e) to 65(1)(j), before the employer can give notice of termination the employer's representative must certify in writing to the employer, with a copy to the contractor, that in his opinion the contractor is in default.

The question arises, can this be challenged and what happens if it is? Can the employer enter the site and expel the contractor or must he await the outcome of arbitration proceedings?

These were amongst the questions considered by the court in the *Tara Civil Engineering* case under an ICE Fifth edition contract. The engineer certified that Tara was in default and the employer gave notice of its intention to expel Tara from the site. Tara obtained an ex parte injunction restraining the employer from expelling them to which the employer responded by applying to the court for the order to be discharged. Granting the order requested by the employer, the court held that it should not go behind the engineer's certificate under clause 63 or any other documents relied on unless there was proof of bad faith or unreasonableness. The judge said:

> 'The most important of the three documents setting clause 63 in motion is the certificate of the engineer as to his opinion. It is important that the certificate is of his opinion only and not of fact. I take the view that I should only go behind that certificate, or behind any of the other documents relied on as setting in motion the clause 63 procedure, if there is either a lack of documents which on their face appear to set the procedure in motion or there is proof of bad faith or proof of unreasonableness.'

And later in the judgment, commenting on the policy of the courts, he said:

> 'At this stage there is no intention by the court to take sides in the determination of the ultimate disputes between the parties. The concern of the court is far from seeking to assist either party to break the contract. It is impossible to decide at this stage what is the conduct which would be in breach of the substantive terms of the contract. The court's present concern is to enforce the terms of the contract with regard to the only matters presently under consideration, namely, the regulation of the conduct of the parties pending the resolution of the substantive dispute by arbitration.'

The employer can, therefore, act on the certificate of the employer's representative, even if it is challenged. But, of course, the employer will be liable if an arbitrator later overturns the certificate.

Completing the works

Clause 65(2) provides that where the employer has entered the works after expelling the contractor, the employer may:

- complete the works himself
- employ another contractor to complete the works.

The employer can use for completion so much of the contractor's equipment and temporary works as is deemed to be the employer's property under clause 54. And he can sell any of the contractor's equipment, temporary works, unused goods and materials and apply the proceeds towards the satisfaction of sums due to him from the contractor.

These are greater powers in theory than in practice. They only work to the full if the contractor has title to all the equipment, temporary works and unused goods and materials. He may, in fact, have little title, with most equipment and temporary works on hire and most unused goods and materials not paid for.

Assignment to the employer

Clause 65(3) empowers the employer's representative to require the contractor to assign to the employer the benefit of any agreement for the supply of goods, service or execution of work.

It is not wholly clear what is meant by assigning the benefit of any agreement. It cannot be thought that any supplier or sub-contractor will provide goods or services to the employer with the burden of payment left with the contractor.

Almost certainly if any assignment is going to work in practice the employer will have to take on both the benefits and the burdens. And since the burdens will include payment of sums outstanding the employer will usually be better off simply by reaching new agreements with suppliers and sub-contractors.

Valuation at date of termination

The employer's representative is required by clause 65(4) to value the work at the date of termination by fixing and certifying:

- the amount earned by the contractor in respect of work done, and
- the value of unused goods and materials and any contractor's equipment and temporary works deemed to be the property of the employer under clause 54.

The purpose of this task is not obvious. The amount which may eventually become due to the contractor is arrived at by the method of calculation given in clause 65(5) and the valuation at the date of termination is not included in the calculation.

Payment after termination

Clause 65(5) sets out a scheme where no payments are due from either party after termination until the expiration of the defects correction period or such later date as all the employer's expenses have been ascertained and the amount certified by the employer's representative.

Similar provisions have been criticised as delaying the employer's right to recovery of sums due. In many cases the employer, by letting a new contract, will know well before completion what his loss is. With a common law determination his entitlement to recovery of loss would be immediate. Under these provisions the employer is caught with the contractor in what could be a prolonged wait for settlement.

Sums already certified

The employer can rightly refuse to pay on interim certificates coming due for payment after termination. Even if there are certificates which are overdue, as frequently happens when termination is anticipated and payments are held back, the employer should still be able to mount a successful counterclaim against any writ for payment.

Damages for delay

There is no certainty on the wording of clause 65(5) whether the damages for delay in completion are liquidated damages accrued to the date of termination plus general damages thereafter or liquidated damages for the full period to completion. It is suggested that the latter, which is more consistent with the provisions of clause 47 (liquidated damages), is probably correct.

Value of bonds

It is not clear from clause 65(5) how any sums the employer obtains from security bonds are to be taken into account in settling the final sums due. The sums due are stated only in terms of amounts due on the contract, the costs of completion and expenses.

The employer's representative probably has no power to

consider payments made under bonds. In any event, such payments are often long delayed as bondsmen not infrequently mount a thorough examination of their liability under the exact terms of the bond.

Chapter 20

Settlement of disputes

20.1 Introduction

'I cannot imagine a civil engineering contract, particularly one of any size, which did not give rise to some disputes. This is not to the discredit of either party to the contract. It is simply the nature of the beast.'

Lord Donaldson
The Master of the Rolls.

For employers a big attraction of design and construct contracting is the reduced scope for contractors' claims and hopefully a change in the nature of the beast. Without the engineer to blame for his problems the contractor loses a prime source of complaint.

This should lead naturally to less disputes but other factors may come into play. Lump sum contracts lack the flexibility of remeasurement contracts in providing the contractor with remuneration for additional costs – a flexibility sometimes stretched to the limit to avoid dispute. And contractor's design prevents the employer's representative from devising cost saving solutions to practical problems. Nevertheless there is anecdotal evidence from the building industry that there are fewer formal disputes in design and build than in traditional contracting. It will certainly be a severe disappointment to employers in the civil engineering industry if design and construct contracting fails to prove less contentious than the traditional system.

Claims, differences and disputes

Not every claim leads to a dispute; and not every difference leads to arbitration. Much depends on the capacity of the contract to cope with the legitimate concerns of the parties. Some contracts are remarkably free from dispute – the IChemE model forms in use for 25 years have yet to trouble the courts (recognition, perhaps, of the generosity of their provisions towards the contractor) – but others which give the contractor a harder deal are notoriously dispute prone.

If the contract allows for instructions and orders of the contract administrator and his assistants to be challenged and reversed; for the contractor to be consulted on the issue of variations and their pricing; for contentious matters to be referred to an independent adjudicator or expert, then, disputes in the formal sense should largely be avoided.

Disputes in design and construct contracting

In design and construct contracting it is more important than in traditional contracting that claims and differences should be settled without recourse to formal dispute procedures. Arbitration is a poor solution to a problem which needs immediate attention and decision. And whereas in traditional contracting the engineer can direct and the contractor must comply – thereby ensuring that the works are built – in design and construct contracting an impasse can develop without an alternative mechanism for solving problems.

Consider for example a contractor who under clause 8(2)(b) of the D and C Conditions refuses to accept responsibility for the employer's design unless major changes are made. Suppose the employer's representative refuses to consent. What then? Short of abandoning the works the only solution appears to be conciliation and arbitration – both fairly lengthy processes.

Clearly what is needed in design and construct contracting are procedures for prompt decision making to keep the works progressing leaving arbitration as the longstop remedy for the financial consequences.

Procedures in standard forms

Most of the standard forms of contract currently used for design and construct/design and build have such procedures. For example:

- JCT 81 – provision is made for an adjudicator to deal with disputes arising prior to practical completion or relating to

termination. Ideally the adjudicator is named in the appendix before the contract is signed. Where adjudication applies it is a formal condition precedent to arbitration.
- The Department of Transport Design and Construct form – various options apply but one is that a Disputes Resolution Panel is maintained from the start of the contract until the issue of the maintenance certificate. A unanimous award of the panel is final and binding on the parties. If either party wishes to contest a non-unanimous award it is a condition precedent to arbitration that attempts for amicable settlement are made involving the chief executive officer of the contractor and a designated officer of the department.
- The IChemE forms – certain disputes relating to variations, tests, certificates and termination are expressly referred to an 'Expert'. The parties agree to be bound by any decision of the expert and not to question the decision of the expert in any proceedings.
- The New Engineering Contract – an adjudicator, agreed by the parties or otherwise appointed, is empowered to rule on disputes and assess the financial consequences. Arbitration can follow.
- GC/Works/1 (edition 3) – Single Stage Design and Build – an adjudicator, selected rather unusually by the employer, hears representations and gives decisions which are binding until completion, alleged completion or abandonment of the works. Again, arbitration can follow.

20.2 *Procedure in the Design and Construct Conditions*

The D and C Conditions have not followed the route, illustrated above, of appointing an independent person or panel to settle disputes on an interim, or possibly final, basis during the construction of the works.

Instead the Conditions rely on a more formal three stage process:

(1) If the contractor is dissatisfied with any instruction etc of an assistant he is entitled to refer the matter to the employer's representative for a decision.
(2) If either party is dissatisfied with a decision of the employer's representative or the parties are otherwise in dispute, either party can refer the matter to conciliation.
(3) The conciliator's recommendation is final and binding on the parties unless either party gives notice within three months of

referral to arbitration. However such referral cannot take place prior to conciliation.

Comparison with ICE Sixth edition

The procedure for the settlement of disputes in the D and C Conditions retains much of the detail found in the Sixth edition: Clause 66 has been transposed word for word in many of its paragraphs.

Nevertheless there are two important differences:

(1) There is no provision for the employer's representative to give clause 66 decisions corresponding with the familiar clause 66 decisions of the engineer.
(2) Conciliation is not optional prior to arbitration; it is mandatory and a firm condition precedent to arbitration.

The implications of these changes on the application of clause 66 are examined in detail later in this chapter but there is a wider point to consider. Who controls the execution of the works – the contractor or the employer through his representative?

Control of the works

In the Sixth edition, as in the Fifth edition before it, the scheme for control of the works is clear enough. The engineer designs and the engineer decides. The contractor does as he is told. If the contractor seeks extra payment for what he has been told to do he can go to conciliation or arbitration and obtain whatever recompense is due but he must comply with the engineer's instructions (subject always to considerations of safety).

In the D and C Conditions, the employer's representative has no obvious power to control the works. Clause 13 requiring the contractor to comply with his instructions is omitted and variations can only be ordered on a consultative basis. Consequently in the event of dispute rapid recourse to conciliation or arbitration may be the only way to maintain progress. This throws an interesting light on the role of the conciliator.

Role of the conciliator

In the Sixth edition the role of the conciliator is principally to settle monetary disputes. In the D and C Conditions he is likely to acquire a more active and more practical role. In short a role not much different from that of the adjudicator, expert or panel found in other contracts.

The big difference is that the conciliator is appointed after disputes have arisen and not before; the big disadvantage is that the conciliator's recommendation can take up to three months to be given.

Policy of the Conditions

If it is the policy of the Conditions that the conciliator should replace the engineer in the decision making process it would be better for both parties if he was nominated at the outset to avoid delays. If it is not the policy that the conciliator should have an active role then the authority of the employer's representative needs to be expressed with more force and certainty.

20.3 *Conciliation generally*

The D and C Conditions refer specifically to the ICE Conciliation Procedure (1988) or any later modification. The detail of this procedure is examined in the next section but as it stands the ICE Procedure is some distance apart from conciliation as practised within the realm of alternative dispute resolution.

Alternative dispute resolution

Various techniques for resolving disputes, developed principally in the United States of America, to avoid the high legal costs, acrimony and time consumption of formal proceedings are now gaining ground in this country. These techniques include mini trials, conciliation and mediation. Collectively they are known as methods of alternative dispute resolution (ADR).

The mini trial

The mini trial, or executive tribunal as it is sometimes called, is a non-binding procedure at which the parties present their case before senior executives from both parties assisted by a neutral expert. The role of the expert who may have a legal or technical background is to help the parties focus on critical issues and if appropriate to offer advice on the merits or likely outcome of the dispute if a settlement is not reached.

The disputes resolution procedure in The Department of Transport design and construct form is along the lines of the mini trial and if it is successful it may well set the model for other forms.

Conciliation and mediation

The procedure whereby a neutral person assists the parties to reach a settlement to their dispute can be known as either conciliation or mediation. The terms are not fixed and there is a good argument that flexibility should be maintained in the use of the terms so that the various procedures they encompass can converge and adapt to suit particular circumstances.

Nevertheless there is a move, at least in this country, to ascribe the terms conciliation and mediation to distinct procedures.

This is how the Master of the Rolls, Lord Donaldson, in an address in 1991 to the London Common Law and Commercial Bar Association explained the distinction:

> 'Conciliation: This is a process whereby a neutral third party listens to the complaints of the disputants and seeks to narrow the field of controversy ... He moves backwards and forward between the parties explaining the point of view of each to the other. He indulges in an onion peeling operation. He peels off each individual aspect of complaint, inquiring whether that aspect really matters. In the end he and the parties are left with a core dispute which, so much having been discarded under the guidance of the conciliator, at once seems more capable of settlement on common sense lines.
>
> Mediation: This is what a PR man would describe as "conciliation plus". The mediator performs the functions of a conciliator, but also expresses his view on what would constitute a sensible settlement. In putting forward his suggestion, which the parties will be free to reject, the mediator will in most cases be guided by what he believes would be the likely outcome if no settlement was reached and the matter went to a judicial or arbitral hearing.'

In short, the neutral person takes a more interventionist or positive role in mediation than in conciliation. Sometimes the terms evaluative mediation (for mediation) and facilitative mediation (for conciliation) are used.

Agreement not enforcement

Both conciliation and mediation are voluntary processes and neither produces, of its own accord, a legally enforceable outcome.

The same is true of mini trials, adjudication and experts. Contracts may state that the decisions produced are final and binding on the parties but how effective is this if the decisions cannot be legally enforced?

The courts have not yet been asked to rule on the enforcement of decisions under the newer methods of alternative dispute resolution but there have been a few cases on adjudication. In *A. Cameron Ltd* v. *John Mowlem & Co.* (1990) the Court of Appeal ruled that an adjudicator's award could not be enforced as an arbitrator's award, since his decision could not be regarded as 'award on an arbitration agreement' within the meaning of Section 26 of the Arbitration Act 1950.

Enforcement of two stage procedures

However the courts have indicated, most recently in the case of *Channel Tunnel Group Ltd* v. *Balfour Beatty Construction Ltd* (1993) that they will stay proceedings to support alternative dispute resolution procedures. Eurotunnel argued that the court could not stay the matter to arbitration because it was a panel of experts to whom the matter must first be sent. Lord Mustill granting the stay said:

> '...I believe that it is in accordance, not only with the presumption exemplified in the English cases ... that those who make agreements for the resolution of disputes must show good reasons for departing from them, but also with the interests of the orderly regulation of international commerce, that having promised to take their complaints to the experts and if necessary to the arbitrators, that is where the appellants should go.'

Compulsory or optional ADR

Arbitration and litigation produce legally binding and enforceable decisions. In short, they produce a result. Alternative dispute resolution cannot guarantee to achieve this. Normally there is no binding decision and even if the decision is said to be binding there is the question of enforceability.

When alternative dispute resolution is being used as originally conceived, with willing parties to a voluntary arrangement, such matters are secondary to the benefits both parties perceive to exist. However, with compulsory ADR and an unwilling party on one side, clear disadvantages begin to emerge:

- it may simply be a time wasting exercise
- it may be used solely to assess the strength of the opposition case
- it may complicate later proceedings by dispute on the admissibility of evidence and the extent of 'without prejudice' disclosures.

Perhaps it amounts to this – compulsory conciliation is a contradiction in terms.

20.4 The ICE Conciliation Procedure (1988)

The ICE Conciliation Procedure was first seen in the ICE Minor Works Conditions of Contract in 1988 – setting a new trend for the construction industry. Since then the Procedure has been made optional in the Sixth edition and compulsory in the D and C Conditions.

The procedure is not obviously conciliation or mediation as defined above – it looks more like an informal version of Part F, the Short Procedure, of the ICE Arbitration Procedure.

However the training for the ICE list of conciliators encompasses both the conciliation and mediation techniques described earlier so a flexible approach to conciliation may well be encountered in practice.

Rules of the ICE Procedure

The principal rules of the ICE Procedure as it presently stands are as follows:

- a conciliator to be agreed within 14 days of notice to refer being given;

- the conciliator to start as soon as possible and to use his best endeavours to conclude as soon as possible – and in any event within two months of appointment;
- the parties to send to the conciliator and each other written submissions of their case with copies of all relevant documents relied on;
- within 14 days of written submissions, further written submissions by way of replies to be made;
- the conciliator may inspect the site or otherwise inform himself of the facts of the dispute 'including meeting the parties separately';
- the conciliator may convene a meeting to take evidence and hear submissions but is not bound by the rules of evidence or any rules of procedure;
- within 21 days of the conclusion of the meeting the conciliator is to prepare his recommendations as to the way the matter is to be settled;
- the conciliator is to notify the parties that his recommendation is prepared and upon receipt of his fees is to issue his recommendation to both parties.

Nature of the Procedure

There can be little doubt that as currently drafted the ICE Conciliation Procedure indicates a merit based solution – even if there is some departure from the rules of natural justice in permitting the conciliator to meet the parties separately.

However there is a view that conciliation, even when commenced under the ICE Procedure, should be conducted as mediation with the conciliator aiming to produce not so much a recommendation based on merit as an equitable solution the parties can live with. To this end it is expected that the 1988 Procedure will be replaced in due course with more flexible rules.

In the meantime the parties do need to know what they have let themselves in for. Are they presenting their case in expectation of an independent recommendation or are they entering into negotiations to settle. The point is of fundamental importance because neither true conciliation nor mediation can work unless the parties are represented at the conciliation by persons with authority to settle.

Will conciliation work?

If the intention of introducing conciliation into the D and C

Conditions and other ICE contracts is to reduce the number of disputes going to arbitration it should be successful however it is conducted.

But a side effect of this may well be an overall increase in the number of disputes. Once contractors realise they have an opportunity to put any claim before a neutral person for the outlay of modest costs they are likely to become enthusiastic players of the game.

20.5 *Arbitration generally*

Arbitrations in England and Wales are regulated by the Arbitration Act 1950 as amended by the Arbitration Act 1979. Section 32 of the 1950 Act defines an 'arbitration agreement' as a 'written agreement to submit present or future differences to arbitration, whether an arbitrator is named therein or not'.

Enforcing the arbitration agreement

If there is a valid arbitration agreement covering the dispute the courts will normally stay any legal proceedings brought by one party, thus upholding the agreement – section 4 Arbitration Act 1950. The courts will not grant a stay however if the applicant for the stay has taken a step in the legal action, that is, shown himself willing to contest the legal proceedings.

In order to obtain a stay the application must:

- show there is an enforceable arbitration clause in the contract
- make an application at an early stage in the proceedings
- establish that he is ready and willing to arbitrate.

Of course the parties may, if they are of the same mind, avoid the arbitration agreement altogether and go to court.

Multi-party actions

In construction disputes there will frequently be numerous parties involved – contractor, employer, sub-contractors, suppliers, designers. But arbitration, unlike litigation, cannot cope with multi-party disputes unless all the parties to the proceedings so agree or unless the arbitration agreement requires the use of rules of procedure which cater for multi-party actions.

Because different sets of proceedings on the same dispute would involve extra cost, procedural difficulties and a grave risk of inconsistency the courts may exercise their discretion and refuse a stay. Lord Justice Pearson in *Taunton-Collins* v. *Cromie* (1964) said:

> 'Two well established and important principles. One is that parties should normally be held to their contractual agreements ... the other principle is that a multiplicity of proceedings is highly undesirable for the reasons which have been given. It is obvious that there may be different decisions on the same question and a great confusion may arise.'

Scope of the arbitration

The arbitrator's jurisdiction derives from the arbitration agreement and the courts distinguish between disputes arising 'under' a contract and disputes arising 'in connection with' a contract.

Arbitration agreements in modern construction contracts are generally drafted with wide wording to ensure the agreement is not limited in the scope. Thus the D and C Conditions state 'in connection with or arising out of the Contract or the carrying out of the Works'.

The question of whether an arbitrator can consider matters which strike at the very existence of the contract – misrepresentation and the like – has been in debate for many years. The point being that if the arbitrator rules there is no contract he has apparently lost the authority on which to make his ruling.

However in *Ashville Investments* v. *Elmer Contractors* (1987) the Court of Appeal held that claims arising out of alleged innocent or negligent misstatements were claims arising 'in connection with' the contract and the arbitrator had jurisdiction.

Control by the courts

The courts have a general supervisory role over arbitrations and under powers given by the 1950 Act the courts may:

- appoint an arbitrator if the parties cannot agree
- make orders for security and examination on oath

- remove an arbitrator who is guilty of delay, misconduct or bias
- remit an award to an arbitrator for his reconsideration
- set aside an award for misconduct.

Under the 1979 Act the courts may:

- review arbitration awards
- determine preliminary points of law
- extend the time for commencement
- enforce awards.

Appeals to the courts

Section 1(2) of the 1979 Act permits appeals on questions of law arising out of an award with the consent of the parties or with the leave of the court. Leave is not granted unless the court considers that the determination of the question concerned could substantially affect the rights of one of the parties. Generally, leave to appeal will only be granted where there is a standard form of contract involved. The point in issue must be one of public interest. In one-off contracts leave to appeal will only be granted in special circumstances.

Under the 1979 Act the parties can enter into an exclusion agreement which precludes rights of appeal on points of law and frequently do so in short form civil engineering arbitrations.

Conduct of the arbitrator

An arbitrator is bound by two principal rules:

- to act fairly between the parties in accordance with the rules of natural justice
- to decide only the matters submitted to him in accordance with the arbitration agreement.

Failure to act fairly can lead to removal of the arbitrator by the courts for what is termed 'misconduct'. Recent examples include:

- an arbitrator making an interim award on a point of law without hearing one of the parties – *Modern Engineering (Bristol) Ltd* v. C. *Miskin & Sons Ltd* (1981)

- an arbitrator failing to give a party an opportunity to deal with his (the arbitrator's) own special knowledge of the facts – *Fox* v. *P. G. Wellfair Ltd* (1981)
- an arbitrator receiving a prejudicial document from one party and refusing to let the other party have a copy of it – *Maltin Engineering* v. *Dunne Holdings* (1980).

If the arbitrator exceeds his jurisdiction the courts will set aside his award as in *Secretary of State for Transport* v. *Birse-Farr Joint Venture* (1993).

Characteristics of arbitration

Attempts are frequently made to list the advantages and disadvantages of arbitration compared with litigation. The exercise is largely pointless because individual experiences will always suggest different conclusions.

However, arbitration does have characteristics which distinguish it from litigation, some of which may be of advantage; some of disadvantage. It depends upon what the parties are seeking.

The principal characteristics are as follows:

- the hearing is private
- the award is published privately
- the parties may be able to choose their own arbitrator
- the venue of the hearing is flexible
- the hours, date and all administrative details of the hearing are flexible
- the hearing can be less formal than in litigation
- the rights of appeal against an award are restricted
- not ideal for multi-party disputes
- the arbitrator and the venue have to be paid for
- lawyers are optional
- lay advocates are allowed
- the arbitrator has no general power to order security for costs but may have under some rules
- a reluctant party may be less robustly dealt with
- can be cheaper and quicker
- documents-only may be a permitted procedure
- the arbitrator has power to award interest and costs
- the arbitrator can make site visits etc
- the arbitrator may be given powers the courts do not have

- arbitration may be imposed if an agreement is included in the contract
- the legal aid scheme is not applicable

20.6 The ICE Arbitration Procedure (1983)

Clause 66(7)(a) of the D and C Conditions requires that any reference to arbitration shall be conducted in accordance with the ICE Arbitration Procedure (1983) or any amendment or modification thereof in force at the time of appointment of the arbitrator. The ICE Procedure applies also to the ICE Fifth and Sixth editions and to the ICE Minor Works Conditions.

Rules of the Procedure

The aim of the Procedure is to bring out and develop those features of arbitration which distinguish it from and give it an advantage over litigation. It does this with great success. In leaflet form the Procedure details rules on:

- notices for referral
- appointment of the arbitrator
- scope of the arbitrator
- powers of the arbitrator
- procedures at the hearing
- pleadings and discovery
- procedural meetings
- preparation for the hearing
- summary awards
- evidence
- the award
- reasons
- appeals
- the Short procedure
- the Special procedure for experts
- interim arbitrations.

Although many of the points covered are simply powers and procedures which derive from the Arbitration Acts, the extended powers of the arbitrator and the Short and Special procedures make the ICE Procedure (1983) by far the best set of rules for construction arbitrations.

Powers of the arbitrator

The express powers given to the arbitrator in the ICE Procedure include:

Rule 6 – power to order protective measures for the subject matter of the dispute
– power to order deposits of money for security of costs
Rule 7 – power to order concurrent hearings, with the agreement of all parties, if the dispute concerns two or more contracts
Rule 9 – power to order legal, technical or other assessors to assist in the arbitration
– power to seek legal or technical advice
– power to rely on his own knowledge and expertise
Rule 11 – power to order pleadings or statements
– power to order discovery of documents
Rule 14 – power to make summary awards
Rule 15 – power to proceed ex parte
Rule 16 – power to order disclosure or exchange of proofs of evidence
– power to put questions and to conduct tests.

Procedural options

The ICE Procedure gives rules in detail for three methods of conducting arbitrations:

- Parts C, D and E – full procedure with hearing
- Part F – the Short procedure
- Part G – the Special procedure for experts.

The Procedure also states that the arbitration may proceed on documents-only with the agreement of the parties. There are no detailed rules for this but in practice it is a variation of the Short procedure.

The decision on the method to be adopted is essentially a matter of agreement for the parties although usually there will be discussion between the arbitrator and the parties at the preliminary meeting on the best way to proceed. However the rules preserve the right of a party to be heard so the arbitrator cannot embark on the use of the Short procedure, the Special procedure or documents-only without the agreement of the parties. There is an exception to this in Part H in respect of interim arbitrations – which are arbitrations conducted prior to completion of the works.

Two points which frequently influence the parties in selecting a particular procedure and which often go against the use of the Short and Special procedures are discovery of documents and awards of costs. Neither of these procedures involves discovery and neither permits the arbitrator to make any award on legal costs unless the parties so agree. However, the arbitrator may suggest to the parties variations of the Short and Special procedures to overcome these difficulties.

Part F – Short procedure

The Short procedure is appropriate for disputes on single issues, measurement claims, interpretation of documents and other like situations where a fair decision can be reached without hearing factual or expert evidence.

Although the rules only require exchanges of statements of case the arbitrator will probably give directions for the service of replies. This ensures that each party is better prepared for oral submissions and it assists the arbitrator to identify specifically those points on which the parties are in agreement or disagreement.

The rules require the arbitrator to convene a meeting with the parties within a month of exchange of statements and associated steps to receive oral submissions and to put questions. Such meetings are relatively informal compared with a full hearing and whilst the parties are free to appoint lawyers as their advocates many hearings are conducted with lay advocates on both sides.

Part G – Special procedure

The Special procedure for experts is based on the Short procedure but is designed to apply to limited issues which depend upon expert evidence. After receipt of each party's file and after viewing the site if he thinks it necessary, the arbitrator meets the expert witnesses who may address him and question each other.

Part H – Interim arbitration

Rule 24 of the Procedure states that if the arbitration is to proceed before completion, or alleged completion of the works, it shall be called an interim arbitration.

The rule is clearly intended to apply to disputes which need quick resolution to facilitate completion of the works but the wording makes it of general application to disputes before completion whether or not there is a matter of urgency.

Under rule 24.7 the arbitrator is given power to direct the use of Part F (the Short procedure) or Part G (the Special procedure) which does reduce the rights of the parties. This difficulty, if it is seen as such by the claiming party, can of course be avoided in the case of non urgent disputes simply by deferring action under clause 66 until after completion.

In an interim arbitration the arbitrator can make:

- final awards
- interim awards
- findings of fact
- summary awards
- interim decisions.

Anything which is not expressly identified as one of the above is deemed to be an interim decision. The significance of this is that an interim decision is final and binding only to completion and can thereafter be re-opened in arbitration before the same or a different arbitrator.

20.7 Clause 66 of the Design and Construct Conditions

Clause 66(1) – settlement of disputes

Clause 66(1) defines in broad terms what constitutes a dispute and provides that any dispute shall be settled in accordance with the provisions of clause 66. That is to say, by conciliation and arbitration.

A dispute can only exist between the parties to the contract but such dispute can include any dispute or difference as to a decision, instruction, certificate etc of the employer's representative.

The opening words of the clause 66 'Except as otherwise provided in these Conditions' refer most obviously to clause 10 (performance bonds) and clause 70 (value added tax) which have separate rules for dispute settlement. The words may also be of more general application to referrals to the employer's representative under clause 2 and elsewhere.

Clause 66(2) – notice of dispute

Clause 66(2) places disputes on a formal footing by requiring that one party should serve on the other a notice in writing. That fixes when the dispute is deemed to arise for the purposes of conciliation and arbitration.

The provision in clause 66(2) that no notice of dispute may be served unless the party wishing to do so has first taken steps or invoked any procedure elsewhere 'in the Contract' is potentially troublesome.

It could mean that the contractor cannot, for example, raise a notice of dispute on extension of time unless he has made an application under clause 44 or raise a notice on additional payment without claiming under clause 53. It could also relate to clause 2(6) – the contractor's reference on dissatisfaction. It could also apparently relate to matters which are not in the Conditions but are in 'the Contract'.

Whatever the provision means, the problem is in its application. Who decides whether a notice of dispute is validly served? Can there be a dispute about a notice of dispute? Could an arbitrator's jurisdiction or award be challenged on the grounds of invalid notice? No doubt these will all become matters for lawyers to contemplate in due course.

Clause 66(3) – conciliation

The D and C Conditions are unusual in making conciliation a condition precedent to arbitration. In other words, conciliation is compulsory

Clause 66(3) achieves this by stating that every dispute not settled within one calendar month from service of the notice of dispute shall be referred to conciliation.

It is probable that the stipulated time of one month in clause 66(3) is meant to impose a time limit on going to conciliation. That is, unless notice to refer is given within one month the dispute is deemed to be settled and cannot be pursued.

There are, however, objections to this. Firstly literal reading of clause 66(3)suggests that a notice to refer cannot be made until 'after' one month. The second objection is on the deemed to be settled point, for unlike the Sixth edition there is no engineer's decision on the dispute which can be held to be final and binding.

The pointer in favour of the one month being a time limit for referral is found in the last sentence of clause 66(3). That sentence says the conciliator shall make his recommendation in writing and shall inform the parties within three calendar months of the service of the notice of dispute. That is to say the conciliation should be completed within three months of the notice of dispute.

However, even with a one month time limit on referral this will present some difficulty. The ICE Conciliation procedure allows 14 days for appointment of the conciliator and two months from appointment to conclusion of the conciliation. To be certain of obtaining a recommendation within three months of the notice of dispute it would therefore be necessary to refer to conciliation within 14 days of service of the notice of dispute.

The uncertainty of the wording of clause 66(3) on this important matter of timing is unfortunate. It may even provide a means of avoiding conciliation and going straight to arbitration.

Clause 66(4) – effect on contractor and employer

Clause 66(4) states that unless the contract has already been determined or abandoned the contractor shall continue to proceed with due diligence. It further states that the recommendation of the conciliator is final and binding on the parties unless revised by an arbitration.

The first part of the clause seems to be at odds with clause 64(1) in respect of disputes on failures of payment for that clause allows the contractor to suspend or reduce his work rate.

The second part making the recommendation of the conciliator final and binding is effective only in so far as there is a means of enforcing the recommendation. And that, as discussed earlier in this chapter, could present problems.

Clause 66(5) – arbitration

Clause 66(5) permits a reference to arbitration where either:

- a party is dissatisfied with the recommendation of a conciliator, or
- a conciliator fails to give a recommendation within three calendar months after service of the notice of dispute.

The party seeking arbitration has three months from receipt of the conciliator's recommendation or six months after the service of the notice of dispute, whichever is applicable, to refer to arbitration.

The time limits imposed by clause 66(5) are intended to prevent any reference to arbitration later than six months after service of notice of dispute. However as indicated in the comment on clause 66(3) above there may be some uncertainty on their effectiveness if the conciliation is not completed within three months. Indeed, the possibility appears to be open for a reference to arbitration when the conciliator is still operating within the time limits of the ICE Conciliation Procedure. This is unfortunate and it would be better if clause 66(5)(b) said three calendar months after 'reference to conciliation' instead of after 'service of the Notice of Dispute'.

Clause 66(6) – appointment of arbitrator

Clause 66 provides that if the parties fail to agree on the appointment of an arbitrator within one calendar month of service of a 'notice to concur' the dispute shall be referred to a person appointed by the President of the Institution of Civil Engineers.

The Institution maintains its own list of arbitrators for this purpose.

Clause 66(7) – arbitration procedure

Clause 66(7)(a) requires that any arbitration shall be conducted in accordance with the ICE Arbitration Procedure (1983) or any modification in force at the time of appointment of the arbitrator. The clause also provides that the arbitrator shall have full power to open up, review and revise any decision, instruction, certificate etc of the employer's representative.

Clause 66(7)(b) states that neither party is limited in arbitration proceedings to the arguments used in conciliation. Lawyers have been quick to point out the opportunities this gives for using conciliation as a vehicle to test the strengths and weaknesses of an opponent's case.

Clause 66(7)(c) states that the award of the arbitrator shall be binding on all the parties. The choice of the word 'all' may follow the joinder provisions in the ICE Arbitration Procedure but it can apply only to the actual parties to the arbitration and not to third parties.

The 'binding' provision in this clause does not prevent appeal against the arbitrator's award to the courts on a point of law nor does it prevent the award being set aside.

Clause 66(7)(d) permits the arbitration to proceed before completion of the works 'Unless the parties otherwise agree in writing'. That is to say the parties do not have to agree to the arbitration proceeding; they have to agree to it not proceeding. This is in line with the Sixth edition but it is the reverse of the position in many other construction contracts.

The parties may of course refer to arbitration simply to avoid being time-barred and may then postpone commencement of the arbitration itself until after completion of the works.

Clause 66(8) – witnesses

Clause 66(8)(a) confirms that the employer's representative can be called as a witness in either conciliation or arbitration notwithstanding his role as the contract administrator. He can therefore be examined in such proceedings on his decisions and certificates. Normally an engineer or an employer's representative will appear only as a witness of fact but it is not unknown for one to appear as an expert witness.

Clause 66(8)(b) states that all matters and information are placed before a conciliator without prejudice and that the conciliator cannot be called as a witness in arbitration or any subsequent legal proceedings.

This does not mean that documents put forward in conciliation are privileged against discovery in arbitration. It simply means that admissions or concessions made in conciliation cannot be used in arbitration or litigation.

As to the conciliator not being called as a witness in any legal proceedings this suggests that the parties will have to ensure that any agreement reached in conciliation is properly recorded to avoid problems of verification.

These matters are viewed with great interest by construction lawyers as a potential source of fee income and it will indeed by ironic if conciliation procedures which have been designed to eliminate legal costs prove to be yet another source of income.

20.8 Concluding comment

If the movement towards design and construct in civil engineering does no more than reduce the level of claims and disputes it will be adjudged, by employers at least, to be a success. Perhaps in a decade or so the words of Lord Justice Lloyd in the *McAlpine Humberoak* case, 'It seems to be the practice in the construction industry to employ consultants to prepare a claim almost as soon as the ink on the contract is dry' will seem a thing of the past.

If the Design and Construct Conditions contribute towards that end they will have served the industry well.

Appendix

Schedule of comments on drafting of the ICE Design and Construct Conditions

Clause no.	Description	Comment	Section
1(1)(c)	employer's representative	'person' appointed – no reference to firms	7.2, 9.4
1(1)(e)	employer's requirements	'performance and/or objectives' seem to imply fitness for purpose	4.5
1(1)(f)	contractor's submission	covers only modifications and additions agreed 'prior to the award of the Contract'	3.6
1(1)(g)	the contract	does the definition allow for changes in the contractor's submission post-award?	7.3
1(5)	cost	difficult meaning of 'to be incurred'	7.5, 17.8
1(7)	consents	the addition of 'any consent shall not be unreasonably withheld' worth considering	7.10
2(5)(a)	instructions	the contractor only expressly required to comply with oral instructions	9.7
2(6)	dissatisfaction	having regard to the wording of clause 66 it might be better to say 'shall' refer rather than 'shall be entitled' to refer	9.7
3	assignment	why should either party be put under the pressure of the phrase 'consent shall not unreasonably be withheld'?	10.6
4(4)	sub-contracting	is there a drafting slip in line 2? Neglects 'of' any sub-contractor would read better than neglects 'or'	10.7

Clause no.	Description	Comment	Section
5(1)(d)	ambiguities in contractor's submission	there is a case for giving the employer the right to select between discrepant items	3.6
6(1)(b)	delay in issuing further information	delay not specifically mentioned in connection with extension of time	4.7
6(2)(a)	design and drawings	the phrase 'will comply' taken with the definition in clause 1(1)(e) suggests fitness for purpose	4.5
6(2)(b)	contractor's designs and drawings	should it be the contractor's problem if his design complies with the employer's requirements but not with 'any other provision of the Contract'?	4.7
6(2)(b)	modifications to contractor's designs	can modifications be brought within the contract as defined in clause 1(1)(g)?	3.6, 4.7, 8.2
6(2)(c)	modifications to contractor's designs	as above plus the problem of resolving disagreement on consent	4.7
6(2)(d)	delay in consenting to modifications	delay not mentioned in connection with extension of time	4.7
8.1	contractor's general obligations	how is the proviso 'legally or physically impossible' to apply?	4.7, 8.2
8(1)(a)	contractor's design obligation	is the contractor's obligation in fact to 'complete the design'?	4.2

Clause no.	Description	Comment	Section
8(2)(a)	contractor's design responsibility	if skill and care is intended to exclude fitness for purpose it should be more positively expressed	4.5
8(2)(b)	contractor's responsibility for employer's design	are these intended to be post-award obligations or deemed provisions? What is the position if agreement is not reached on modifications?	4.7
8(3)	quality assurance	questionable involvement of the employer's representative in approving the contractor's quality assurance system	8.4
8(5)	stability and safety	unless this clause is qualified clause 12 prevails	8.5
10(1)	performance security	no express sanction if the contractor fails to provide a bond	8.7
11(1)	provision of site information	'deemed' provision not what is intended	8.8
12(1)	adverse physical conditions	the clause as worded is not confined to ground conditions; 'physical conditions' is too wide	8.9
15(2)	contractor's representative	no express obligation on the contractor to appoint a representative	7.2, 9.9
18(1)	boreholes	need for permission to take necessary boreholes is questionable	8.12
20(1)(a)	care of the works	contractor's responsibility extends to the date of issue of a certificate of substantial completion. What if the issue is delayed?	11.4

Clause no.	Description	Comment	Section
20(2)	excepted risks	the definition in terms of apportioned loss or damage is questionable	11.4
20(2)(b)	excepted risks	what is intended by 'design of the Works for which the Contractor is not responsible'?	11.4
22(2)(d)	exceptions	as above plus the possibility that all the contractor's design is an exception	11.6
26(1)	giving of notices	why no mention of the proviso in sub-clause 26(3)(c)?	12.2
26(2)	repayment of fees etc	appears to apply widely to rates payable by the contractor, and to fees for statutory checks on design	4.7, 12.2
26(3)(a)	contractor to conform with statutes	perhaps better as the unavoidable result of complying with the employer's requirements	12.2
28(2)	patent rights	how widely can compliance with a 'specification' apply?	12.3
29(1)	interference	consider the implications of the 'Contract' including the contractor's submission	12.4
30(3)	transport of materials	should the employer's representative be acting as a loss adjuster?	12.5
36(3)(a)	checks and tests	appears to apply only if the contract requires quality assurance; but cannot be so intended	13.3

Clause no.	Description	Comment	Section
36(5)	cost of samples	cost might easily fall on the employer	13.3
36(7)	cost of tests	cost might easily fall on the employer	13.3
39(c)	removal of unsatisfactory work	note unusual clause reference. No mention of specified times for removal as in clause 39(a)	13.6
41(1)	commencement date	a requirement for the commencement date to be fixed in writing would be useful	14.2
42(1)	possession of site and access	the opening words 'The Contract may prescribe' may be too wide by including the contractor's submission	14.4
44(1)	special circumstances	may not include acts of prevention by the employer	14.6
44(2)(b)	assessment of delay	should this be limited to the circumstances listed in sub-clause 44(1)?	14.6
44(3)	interim extension of time	refusal of application not expressly stated to be in writing	14.4
44(5)	final determination of extension	is the 14 days for issue compatible with the 28 days allowed for application?	14.6
46(3)	accelerated completion	the employer's representative can request acceleration but only the employer can agree the terms	14.8

Clause no.	Description	Comment	Section
47(6)	intervention of variations	over complex provisions – not now necessary	14.9
48(2)	certificate of substantial completion	'instructions' might be better worded. Entitlement to receive a certificate 'within 21 days of completion' might also be better worded	15.3
48(5)	completion of parts	no requirement on outstanding works or provision of operating instructions	15.3
49(2)	works of repair	intention not clearly expressed	15.4
49(4)	failure to carry out works of repair	does 'work as aforesaid' apply to outstanding work; and can the employer undertake this himself during the defects correction period?	15.4
51	variations	notable absence of any reference to changes in the contractor's submission	16.4
52	valuation of variations	deals only with extra cost and delay and is silent on omissions	16.5
52(2)(d)	valuation by the employer's representative	is this only for the purposes of interim payments?	16.5
60(5)	retention	contractors might wish to see a requirement for retention to be held in trust	18.9
60(7)	interest on overdue payments	drafted to accord with a legal case now superseded	18.10

Clause no.	Description	Comment	Section
62	urgent repairs	no mention of contractor's right to payment for work undertaken	17.7
63(1)	frustration	is it wise to define frustration as here?	19.2
63(4)	payment on abandonment	should the date of issue of the defects correction certificate be rather the date of the certificate?	19.4
64(1)	failure to pay the contractor	appears to apply to alleged underpayment as well as to non-payment	19.5
64(4)(b)	default of the employer	'attempts' to assign is a default justifying termination but clause 3 does not make it a breach	10.6
64(5)	removal of contractor's equipment	requirement to remove equipment overlooks possible extension of period of notice	19.7
65(1)(a)	default of the contractor	as 64(4)(b)	10.6
66(2)	notice of dispute	what are the 'steps' or 'procedure' elsewhere in the contract; and what happens if there is a dispute on them?	20.7
66(3)	conciliation	this reads as though a referral to conciliation cannot be made until at least one month after the service of the notice of dispute. The opposite is possibly intended	20.7

Clause no.	Description	Comment	Section
66(5)(b)	arbitration	time runs from the service of the notice of dispute. It should be from referral to conciliation to avoid overlap of proceedings	20.7

Table of cases

Note: refererences are to chapter sections.
The following abbreviations of law reports are used:

AC	Law Reports Appeal Cases Series
ALJR	Australian Law Journal Reports
All ER	All England Law Reports
BLR	Building Law Reports
CILL	Construction Industry Law Letter
CLD	Construction Law Digest
Const LJ	Construction Law Journal
Con LR	Construction Law Reports
C & P	Carrington & Payne's Reports
DLR	Dominion Law Reports
Exch	Exchequer Reports
Giff	Gifford's Reports
HBC	Hudson's Building Contracts
KB	Law Reports, King's Bench Division
LGR	Local Government Reports
LJQB	Law Journal, Queen's Bench
Lloyd's Rep	Lloyd's List Law Reports
NZLR	New Zealand Law Reports
SALR	South African Law Reports
STARK	Starkies Reports
TLR	Times Law Reports
TR	Term Reports
WLR	Weekly Law Reports

Table of clause references: ICE Design and Construct Conditions of Contract

Note: references are to chapter sections

Clause:

Index

Note: references are to chapter sections.